Lightweight Radiant Heating and Cooling Systems

Lightweight Radiant Heating and Cooling Systems: Thermal Power and Economic Analysis investigates the technology for installing heating and cooling without wet and dry screeds in a horizontal or vertical arrangement. It presents the determination of comfortable conditions for using lightweight radiant heating and cooling systems (LRH/CS) and demonstrates how to conduct an economic analysis of available floor heating systems using multi-criteria decision analysis.

Representing the first part of an ongoing study, which contains a body of general information about available traditional surface heating systems and systems based on LRH/CS, this book presents laboratory tests, thermal power calculations, and analysis of the comfort of use of a lightweight radiant heating system (LRHS) with and without heat-dissipating lamellas. It also contains calculations of thermal efficiency or maximum supply temperature based on specific materials' parameters for different types of flooring cover, such as ceramic tile.

This book is intended for practising engineers and researchers involved in the design and operation of energy-efficient and sustainable HVAC systems.

Jacek Karpiesiuk is an assistant professor at the Faculty of Environmental and Civil Engineering and Architecture at the Warsaw University of Life Sciences, Poland. He received his Ph.D. from the Bialystok University of Technology, Poland. Between 1990 and 1999, he was a teacher at the Building and Surveyor Technical School in Bialystok. In 1993, he obtained building qualifications as a construction manager and supervision inspector, and in 2001, a building expert qualification. From 2001, Dr. Eng. Karpiesiuk has been serving as a technical advisor in Eastern Europe, Bulgaria, Malta, and France to Elektra, a world manufacturer of electric floor heating and anti-icing systems. He is the creator of thermal insulation boards made for surface heating systems, such as Thermopanel, Waterpanel installed without screeds, and Thermopian, a pipe covering system. Dr. Eng. Karpiesiuk has participated in ten scientific conferences as an invited speaker. He is the author of 64 scientific and technical articles, including one original monograph.

Lightweight Radiant Heating and Cooling Systems

Thermal Power and Economic Analysis

Jacek Karpiesiuk

CRC Press
Taylor & Francis Group
Boca Raton London New York

CRC Press is an imprint of the
Taylor & Francis Group, an **informa** business

Designed cover image: Józef Bagiński

First edition published 2027
by CRC Press
2385 NW Executive Center Drive, Suite 320, Boca Raton FL 33431

and by CRC Press
4 Park Square, Milton Park, Abingdon, Oxon, OX14 4RN

CRC Press is an imprint of Taylor & Francis Group, LLC

ISBN: 978-1-041-15352-8 (hbk)
ISBN: 978-1-041-15353-5 (pbk)
ISBN: 978-1-003-67900-4 (ebk)

DOI: 10.1201/9781003679004

Typeset in Times
by codeMantra

Contents

From the Author viii
List of Main Signs xi

Chapter 1 Water Radiant Heating and Cooling Structures 1

1.1 Types and Structures of "Heavy" Heating and Cooling Systems 1
1.2 Types and Structures of "Lightweight" Heating and Cooling Systems 4
1.3 Summary of Water Radiant Heating and Cooling for Different Construction Types 10
Abbreviations 11
References 11

Chapter 2 Electric Radiant Heating Structures 12

2.1 Summary of the Different Types of Electric Radiant Heating Constructions 19
Abbreviations 20
References 20

Chapter 3 General Characteristics of Radiant Heating and Cooling 21

3.1 Efficiency of "Heavy" Water Heating and Cooling Systems 25
3.2 Efficiency of "Light" Water Heating and Cooling Systems 30
3.2.1 Wall-Mounted "Light" Heating and Cooling Systems 30
3.2.2 Underfloor "Light" Heating and Cooling Systems 36
3.3 Summary of the "Heavy" and "Lightweight" Radiant Heating and Cooling Characteristics 46
Abbreviations 47
References 47

Chapter 4 Laboratory Research, Thermal Power Calculations, and User Comfort of a Lightweight Radiant Heating System (LRHS) 51

4.1 Description of Measurement Stands and Measurement Methodology 52

4.2 Methodology for Calculating the Thermal Power and Maximum Temperature in the Heating Pipe of a Lightweight Radiant Heating System with Lamellas 55
4.2.1 Theoretical Method for Calculating the Heat Flux Density, Using the NORDTEST NT VVS 127 Method with Lamellas 59
4.2.2 Experimental Method for Calculating Heat Flux Density with Lamellas 65
4.2.3 The Maximum Temperature of the Medium in the Heating Pipe Using the NORDTEST NT VVS 127 Method with Lamellas 67
4.3 Methodology for Calculating the Thermal Power of a Lightweight Radiant Heating System without Lamellas 70
4.3.1 Theoretical Method for Calculating the Heat Flux Density, Using the NORDTEST NT VVS 127 Method, without Lamellas 72
4.3.2 Experimental Method for Calculating Heat Flux Density without Lamellas 77
4.4 Tabular Summary of the Thermal Power of "Heavy" and "Lightweight" Radiant Heating Systems 81
4.5 Temperature Distributions on the Surface of Lightweight Floor Radiators, with and without Heat-Dissipating Lamellas 84
4.6 Summary of Laboratory Research, Thermal Power Calculations, and Comfort of LRHS Use 87
Abbreviations 92
References 92

Chapter 5 Thermal Power of Various Flooring Types in a Lightweight Radiant Heating System 94

5.1 Output Data with Thermal Power Calculation Procedure 95
5.1.1 Materials and Flooring Cover Used in the LRHS 96
5.1.2 Determination of the Design Thermal Power of an Underfloor Heating Layout with Thin Screeds and Lamellas 100
5.1.3 Determination of the Design Thermal Power of an Underfloor Heating Layout with Thin Screeds *without Lamellas* 116
5.1.4 Determination of the Design Thermal Power of a *Lightweight* Underfloor Heating Layout *without Screed and without Lamellas* 129
5.2 Results Analysis 143
5.2.1 The Influence of Cover Flooring on the Floor's Thermal Power 143

5.2.2 The Influence of Dry Screeds on the Floor Thermal Power 145
5.2.3 The Influence of Aluminium Lamellas on the Floor Thermal Power 147
5.2.4 Influence of Coil Spacing and Supply Temperature on the Floor Heating Power 149
5.2.5 The Influence of Thermal Insulation on the Floor Thermal Power 153
5.3 Summary 153
Abbreviations 155
References 155

Chapter 6 Multi-Criteria Decision Analysis (MCDA) with Cost Analysis of Selected Types of Underfloor Heating 158

6.1 Identification of the MCDA Object 161
6.1.1 The Following Variants Were Taken into Account for Analysis 161
6.2 Input Data Specification 162
6.3 Objective Functions and Input Criteria 170
6.3.1 The List and Description of Criteria 171
6.4 The Set of Criteria with Their Analysis and Verification 174
6.4.1 Explanations for Selected Criteria 174
6.5 The Weights of Criteria with Their Analysis and Verification 177
6.6 Selection of the MCDA Methodology Adequate to the Set of Criteria 180
6.7 Presentation of the Computational Algorithm 181
6.8 Discussion and Interpretation of the Analysis Results 182
Abbreviations 186
References 186

Chapter 7 Usability and Economics of Lightweight Radiant Heating and Cooling System (LRH/CS) 188

7.1 LRH/CS Features Confirmed by Research and Analysis 189
7.2 LRH/CS Directions of Development and Research 190
Abbreviations 191
References 191

Index 193

From the Author

The old Polish saying goes that "necessity is the mother of invention". However, before a discovery is made, an idea is needed. The beginning of research work, ultimately aimed at developing a computational model of a lightweight, layered floor with thermal impact, stemmed from the need to improve the thermal conditions in a small area of an existing, inhabited single-family building. The investor needed to carry out this type of modernisation efficiently and inexpensively, without introducing the so-called wet work technologies. The available technologies were reviewed, and it was ultimately decided to implement type B underfloor heating without a heat-dissipation layer (heating pipes were placed in ready-made grooves recessed into the thermal insulation). A "dry" cement screed was laid over the extruded polystyrene (XPS) thermal insulation, and the heating coil and ceramic tiles were glued in place. In this way, the conditions for efficient implementation of the project were met (there was no need to wait for the concrete mortar to mature). However, a certain downside has been noticed: the "dry" screed system is not very cheap. By reviewing the world literature, solutions for radiant heating without screeds were found, the so-called lightweight floor system.

It turned out that the lightweight radiant heating and cooling system (LRH/CS), mainly underfloor heating, is very popular. The reason is its many advantages and its alignment with the recently popular trend of caring for the natural environment through natural energy sources. This contributed to the search for an innovative, original technical solution that would allow floor heating to be performed faster, safely, and cheaply, while maintaining high quality. To do this and ensure the effectiveness of your effort, combine practical and theoretical activities. Lewin characterised this approach very well in these words: "There is nothing more practical than a good theory." Scientist/priest Franciszek Blachnicki paraphrased this sentiment: "...since the most practical thing is a good theory, it was necessary to create a good theory for practice." It put it into practice by founding the Light-Life Movement in the 1960s, which has been dynamically developing to this day worldwide. With all these elements in mind, research began, resulting in the development of an original technical solution for a lightweight floor without screeds. To accomplish this, numerous comprehensive tests, calculations, and analyses were conducted, both in sanitary engineering and in construction (floor structure strength). This monograph describes collected research, analysis, and calculations exclusively related to sanitary engineering, focusing on the comfort and thermal efficiency (thermal power) of the lightweight radiant heating system (LRHS). It should be mentioned that all this research and these calculations culminated in a doctoral thesis written and defended in 2022 entitled "Calculational Model of a Lightweight Layer Floor Loaded by Temperature". It confirms that a lightweight-heated floor can be safely used under various loads, including non-static ones (temperature effects). The only condition for this is the use of appropriate materials and floor layers, as described in the doctoral thesis.

It is constantly looking for solutions for buildings that ensure both comfortable use and high thermal and cooling efficiency. It would be ideal if heating and cooling were primarily provided by devices powered by renewable energy sources, such as heat pumps, photovoltaics, and others. The LRH/CS without any screeds – both "wet" and "dry" – fits perfectly into this concept. This monograph presents the latest experimental research and numerical and mathematical calculations by scientists from around the world, as well as the author's results on a lightweight heating system, without and with heating lamellas, that improves floor heat distribution. The only standard that provides such calculations, but only with metal lamellas, is the Norwegian national standard. Unfortunately, according to the author of this monograph, this national standard contains an error in the calculation of heat flux density (thermal efficiency). The author corrects this error, indicating its nature, and proves computationally and experimentally (taking into consideration his own and other authors' calculations) that the correction given by him in the formula from the Norwegian standard leads to correct solutions.

In addition, the novelty of this monograph is the development of a methodology for calculating the thermal efficiency (thermal power) of lightweight underfloor heating without the use of lamellas. The Norwegian standard provides heat flux density calculations only for a lightweight heating system with lamellas. However, this monograph also includes proprietary formulas for calculating thermal efficiency without lamellas. They were developed based on experiments and using the calculation process given in the Norwegian standard, after correcting the erroneous formula. Using the calculation methodology outlined in the monograph, it is possible to determine the thermal power required for lightweight underfloor heating and cooling using various materials. The many issues discussed make this book an ideal source of information for investors, designers, and contractors looking for new, ecologically friendly, comfortable, and efficient solutions, not only for radiant heating but also for cooling of all types of buildings. This is even more valuable because LRH/CS can be applied to the floor, wall, or ceiling, allowing the heating or cooling power to be adjusted. An additional factor that facilitates the calculation procedure is the author's spreadsheet, which, after entering the necessary data, calculates the thermal power of a lightweight-heated or -cooled floor, both with and without lamellas. This can be a very useful tool for designers who do not need to perform complex, lengthy mathematical calculations; instead, they can use the formulas provided in this book.

In addition, the monograph is an excellent introduction to ongoing research by scientists in sanitary engineering and construction, as well as a helpful resource for students. Here, they will find basic information on materials science, the construction, and the design of a modern, ecological, lightweight floor system. The current state of research in this field has laid a solid foundation, and it is time to build upon it. Due to the many advantages of this system, it would be worth developing this innovative technology and continuing the research that has been started, e.g. in the direction of:

- checking the strength and thermal power of floors other than ceramic – glued floorboards, vinyl, PVC, and different floors,
- subjecting the vertical LRH/CS system to strength, thermal, and cooling tests,

- carrying out modelling and experiments on other insulating and covering materials in the field of sound and impact acoustics because rigid polystyrene materials (EPS, XPS) are not ideal for eliminating noise transmission,
- checking the system's strength with different types and thicknesses of thermal insulation layers.

These are just a few proposals for continuing research on this attractive ecological heating and cooling system.

The writing of this monograph was made possible thanks to many scientists, engineers, and family members. Let me list some of them according to titular order: Prof. Dr. Eng. Mirosław Żukowski, Dr. Eng., Anna Werner-Juszczuk, and Dr. David Becker from the United States of America for proofreading the English language version, as well as technical employees of the Białystok University of Technology, Elektra Kardo, Sika Poland, and BASF. Thanks to them, we were able to perform many necessary experimental tests. I would also like to thank my wife Agnieszka (Master's Degree in mathematics) for her support with complex calculations and her understanding when I devoted a lot of time to scientific work. I would also like to thank the children, especially M.Sc. Eng. Paweł Karpiesiuk, for preliminary research on the lightweight-heated floor system conducted as part of an engineering thesis in the field of environmental engineering written at the Bialystok University of Technology. Grateful to God and people, I dedicate my research to improving the world and advancing science in the future.

List of Main Signs

CAPITAL LATIN LETTERS

K_{Hb}	equivalent heat transmission coefficient without lamellas
K_{Hl}	equivalent heat transmission coefficient with lamellas
L_{fin}	width of the scattering layer
L_G	the gap between the heat-conducting plates
L_k	estimated width of the horizontal heat-conducting adhesive layer and the tile floor
L_U	the external diameter of the heating pipe
R_{CL}	fictitious thermal resistance of the conductive layer
$R_{con,i}$	resistance for conductive layers carefully shaped and permanently glued with a layer of insulation
R_e	thermal resistance under the layer that conducts heat to the room below or outside
R_{HC}	total thermal resistance between the heat source and the conductive layer
R_i	thermal resistance of the layer conducting heat to the heated room
R_o	thermal resistance of the floor layers above the conductive layer
R_{se}	heat transfer resistance under the floor surface
R_{si}	heat transfer resistance on the floor surface
R'_R	total thermal resistance of the coil wall
$R'_{R,con}$	heat transfer resistance of the coil wall
R_U	thermal resistance of thermal insulation
R'_U	linear resistance
S_{WB}	thickness of the conductive layer (adhesive + flooring)
S_{WL}	thickness of the conductive layer (lamella)
T	spacing of heating pipes (coils)

LOWERCASE LATIN LETTERS

b	groove width
d	groove depth
d_a	diameter of the heating pipe
d_i	thickness of individual floor layers above the conductive layer
d_{ip}	internal diameter of the heating pipe
d_k	thermal insulation thickness
d_o	external diameter of the heating pipe
k_{CB}	thermal conductivity coefficient without the dissipation layer (without lamellas)
k_{CL}	thermal conductivity coefficient of the dissipative layer (lamellas)
k_{fin}	thermal conductivity coefficient of devices
k_k	thermal conductivity coefficient of horizontal layers above thermal insulation
q_i^{max}	maximum heat flux density

q_e experimental heat flux density of the tested radiator without lamellas acc. standards [3]
q_{eb} experimental heat flux density of the tested radiator without lamellas (author's calculations)
q_{el} experimental heat flux density of the tested radiator with lamellas
q_i computational heat flux density
q_B^{max} maximum heat flux density without lamellas
q_i^{max} maximum heat flux density with lamellas
q_{ob} computational heat flux density on the floor surface without lamellas (author's calculations)
q_{ol} computational heat flux density on the floor surface with lamellas

GREEK LETTERS

θ_F^{max} maximum permissible floor surface temperature
$\overline{\theta}_F^{max}$ maximum permissible average floor temperature
θ_H supply temperature to the heating coil
θ_i room temperature
θ_H^{max} maximum temperature of the medium in the heating coil
ℓ characteristic length of the conductive layer (lamella)
ℓ_B characteristic length of the adhesive layer with the floor
λ_i thermal conductivity coefficient of individual floor layers
λ_{ins} thermal conductivity coefficient of thermal insulation
λ_k thermal conductivity coefficient of thermal insulation
λ_R thermal conductivity of the heating pipe wall
λ_{WL} thermal conductivity of the conductive layer (lamella)
λ_s highest thermal conductivity of materials surrounding the conductive layer
λ_{WB} average thermal conductivity weight value of the adhesive layer and the floor
α_i heat transfer coefficient on the floor surface
η. efficiency of the underfloor heating system
ϑ_F average temperature of the tile surface from the measurements
$\overline{\vartheta}_F$ average permissible temperature of the floor with lamellas
ϑ_i air temperature

1 Water Radiant Heating and Cooling Structures

Heating and cooling radiant systems are available in various forms, depending on their location (floor, wall, and ceiling), the heating medium (most often water, air, or electricity), or the type of building partition construction. These various heating forms and sources can be found in national standards, European Union standards, manufacturer catalogues and technical data sheets, online resources, literature, or the legal resources of individual countries. The most commonly known "heavy" systems, along with their various construction types, are listed in this chapter. In addition, the so-called "lightweight" radiant heating systems have been identified, whose designs differ from "heavy" systems primarily by not including external screed layers – whether wet or dry. As the name suggests, they are lightweight due to the elimination of concrete, gypsum, and other screed solutions. A specific feature of "lightweight" systems is that they can be with or without a heat-dissipating layer that conducts heat or cold well, usually made of various types of metal.

There are many types of surface heating and cooling systems covered in the literature, along with various standards, electronic resources, manufacturers' catalogues, and legal regulations governing the subject. The following types can be distinguished depending on the location of the heating or cooling medium in the room:

- Floor
- Wall
- Ceiling
- Mixed

Due to their capabilities, water and electric heating are the most popular. Air heating is less common because this type of heat carrier is less efficient than, for example, water. Mathematical calculations show that when 1 kg of water is heated by 1°C, it produces 4.2 kilojoule (kJ) of heat, while with the same air parameters, we will obtain approximately 1 kJ of heat. This means that to achieve the same thermal power, in the case of air as a heat carrier, the volume of the installation supplying or emitting it must be more than 4 times larger to ensure the appropriate height of the rooms and to allow space for the installation of ducts [1].

1.1 TYPES AND STRUCTURES OF "HEAVY" HEATING AND COOLING SYSTEMS

Depending on the method of placing the heating pipes, the most popular standard in Poland (PN-EN 1264-2: 2008+A 1:2012) [2] distinguishes four types of structures of surface water radiators (A, B, C, and D) (Figures 1.1–1.4), while European

DOI: 10.1201/9781003679004-1

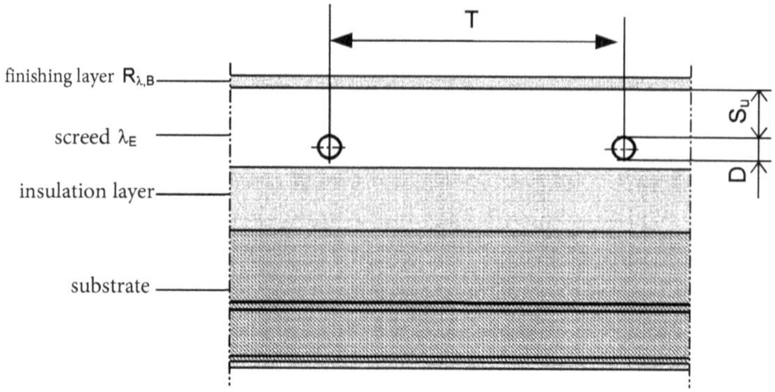

FIGURE 1.1 Heating pipes placed in screed or plaster laid using the wet method over thermal insulation, based on [4], where T – spacing of heating pipes, S_u – thickness of the screed layer above the pipes, D – external diameter of the pipe, $R_{\lambda,B}$ – unit heat conduction resistance of the finishing layer floor, and λ_E – thermal conductivity coefficient of the screed.

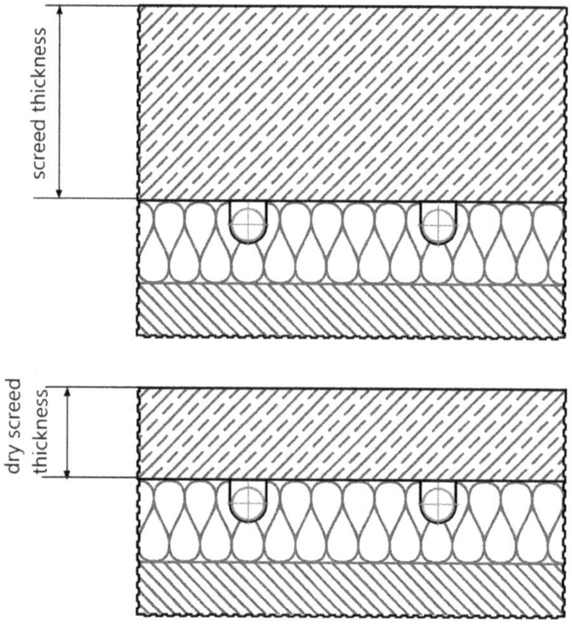

FIGURE 1.2 Heating pipes placed in thermal insulation grooves, based on [5].

standards [3] mention seven types (A–G) (Figures 1.1–1.7). These classifications apply to floor, wall, and ceiling solutions and are as follows:

Type A – Heating pipes placed above the thermal insulation in the screed or plaster layer, made using the wet method from cement or gypsum mortar.
Type B – Heating pipes placed in the upper part of the thermal insulation or the air void, covered with a wet or dry screed made of cement or gypsum boards.

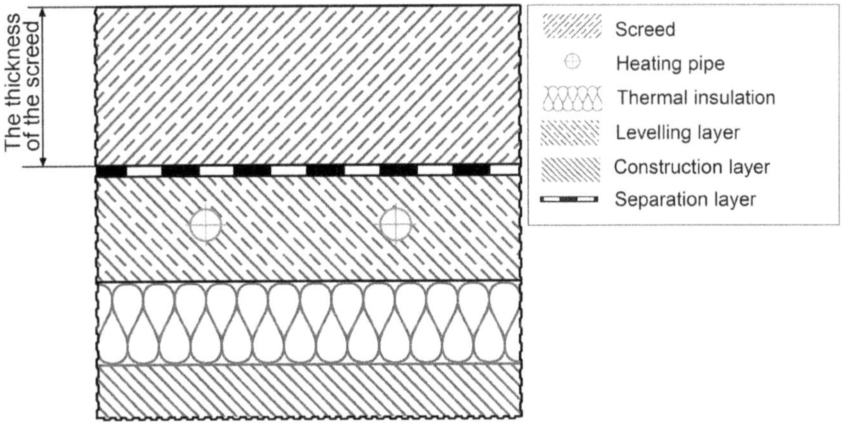

FIGURE 1.3 Heating pipes placed above the thermal insulation in the levelling screed layer, based on [5].

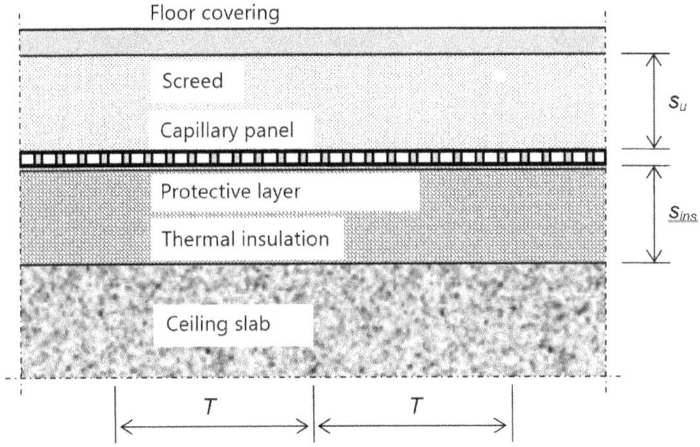

FIGURE 1.4 Adjacent plastic water channels, based on [6].

FIGURE 1.5 Heating pipes arranged in the structural layer, based on [7].

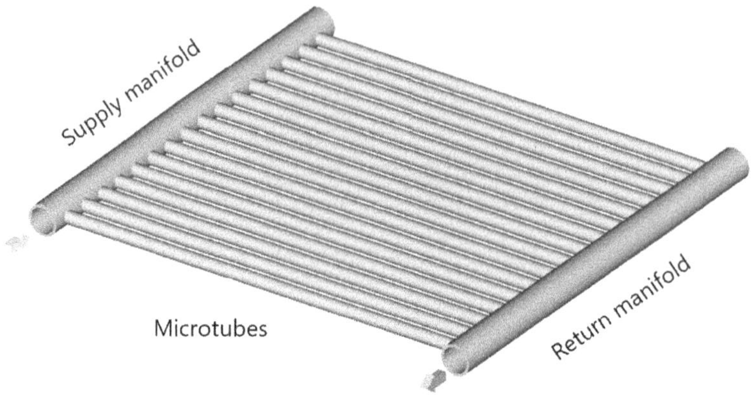

FIGURE 1.6 Small-diameter heating pipes, connected in parallel with collectors, based on [6].

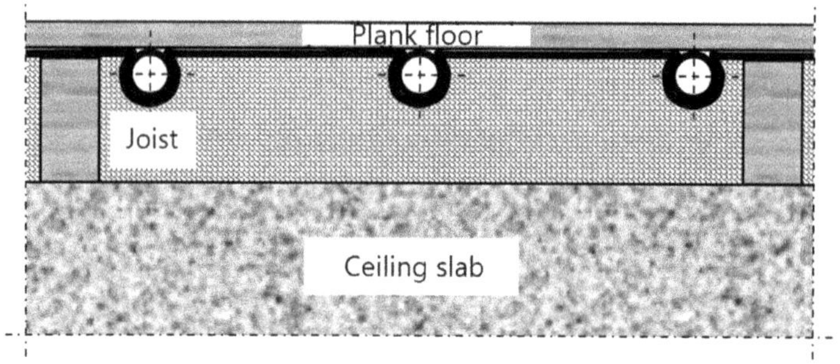

FIGURE 1.7 Heating pipes mounted under the wooden structure between the joists, based on [6].

Type C – Heating pipes placed above the thermal insulation in a levelling screed layer, above which there is a separating layer and another screed.
Type D – Adjacent water channels with a large heat exchange surface, made of low-thickness plastic panels and placed over thermal insulation, covered with a screed or board.
Type E – Heating pipes placed in the structural layer of the ceiling.
Type F – Small-diameter heating pipes, connected in parallel with a supply and return manifold, arranged transversely to the pipes.
Type G – Heating pipes installed under a wooden structure between joists, in a concrete screed or in a structure of profiled steel/aluminium sheets.

1.2 TYPES AND STRUCTURES OF "LIGHTWEIGHT" HEATING AND COOLING SYSTEMS

There are many more solutions to the problem of providing heat to human structures in the world than just those listed in the standards. For instance, there are the so-called "lightweight" radiant heating and cooling structures. Lightweight structures differ from the traditional ones included in standards most frequently, in that, in such systems, the thermal insulation layer is made of materials with relatively low thermal conductivity (wood-like, e.g., chipboard, expanded polystyrene, extruded polystyrene (XPS), gypsum composites, and others of a similar nature). In such a system, heat distribution is supported by a material that conducts heat well from metal sheets or lamellas, most often made of galvanised steel or aluminium. Because of this, the thickness of the radiator structure can be reduced. At the same time, the supply water temperature to the lightweight radiant system must be adjusted, because a layer that conducts heat well not only evens out the temperature distribution on the floor surface but also increases the thermal efficiency of the surface radiator and increases its temperature [5]. Standards PN-EN 1264-2: 2008+A 1:2012 [2] and PN-EN 15377-1:2008 [3] do not apply to lightweight solutions and are supplemented by [8], which are addressed only to light horizontal heating systems. It can also be used when designing electric underfloor heating. Other lightweight structural solutions for radiant heating, without the use of cladding boards or conductive lamellas, can only be designed through experimental tests. Method [8] cannot be used when designing underfloor heating covered with concrete screeds. In addition, to design lightweight heating, the following conditions must be met:

- plate (lamella) must have much higher thermal conductivity than other materials in the structure, which is described by condition (1.1)

$$\lambda_{WL} \geq 10\lambda_s \tag{1.1}$$

λ_{WL} – Thermal conductivity of the conductive layer (lamella) [W/mK],
λ_s – Highest thermal conductivity of materials surrounding the conductive layer [W/mK].

- The product of the thermal conductivity of the conductive plate and its thickness must meet condition (1.2)

$$S_{WL} \cdot \lambda_{WL} \geq 0.01 \tag{1.2}$$

S_{WL} – Thickness of the conductive layer (lamella) [m]

Specific solutions for lightweight structures can, in addition, be linked to the measurements described in Annex A, by the method given in the NT VVS127 standard: UDC 697.1:2001–11 [8] in order to increase the accuracy of the calculations. If the lightweight system is placed directly on the ground, the thermal insulation thickness should conform to the minimum standard requirements in [2].

The standard structural system of a lightweight floor, wall, or ceiling usually consists of the following layers:

- construction base
- vapour barrier, e.g., polyvinyl chloride foil
- thermal/acoustic insulation
- heat-conducting layers
- plates, cables, foils, or heating pipes
- floors or cladding laid loose or glued

An example of a cross-section of such a solution is provided by Karpiesiuk and Chyży in [9], based on research, as shown in Figure 1.8a and b.

The above layering arrangement may vary slightly, depending on the system.

The most popular designs of light surface water heating and cooling are shown in Figures 1.9–1.17.

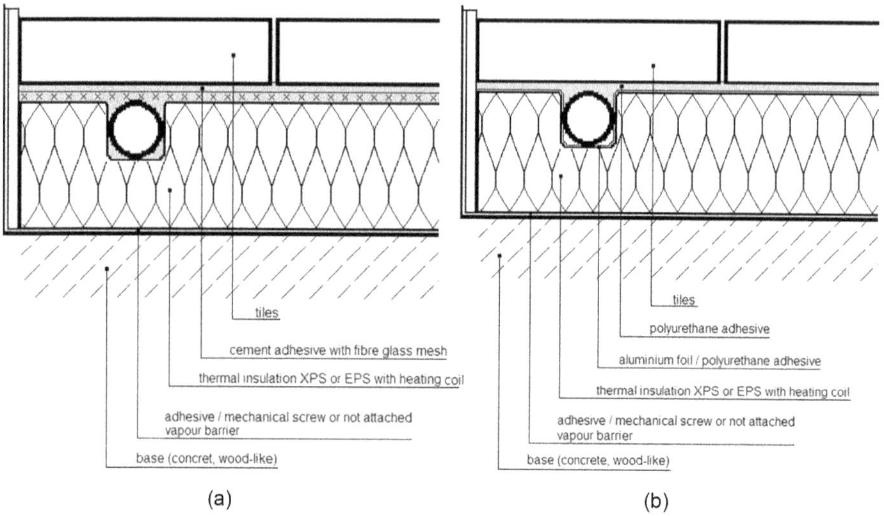

FIGURE 1.8 Cross-sections of a light surface radiator: (a) without the use of a heat dissipation layer (lamellas) and (b) using heat-dissipating lamellas.

FIGURE 1.9 Installation of wall heating with plasterboards from [10].

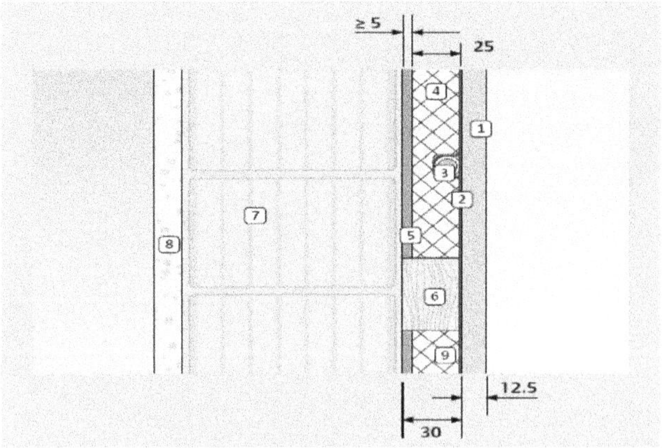

FIGURE 1.10 The vertical cross-section through the external wall of a building with thermal insulation from [10]: 1 – plasterboard, 2 – radiating lamellas, 3 – coil, 4 – system thermal insulation, 5 – additional thermal insulation, 6 – wooden block assembly, 7 – external wall structure, 8 – external thermal insulation walls, and 9 – vapour barrier foil.

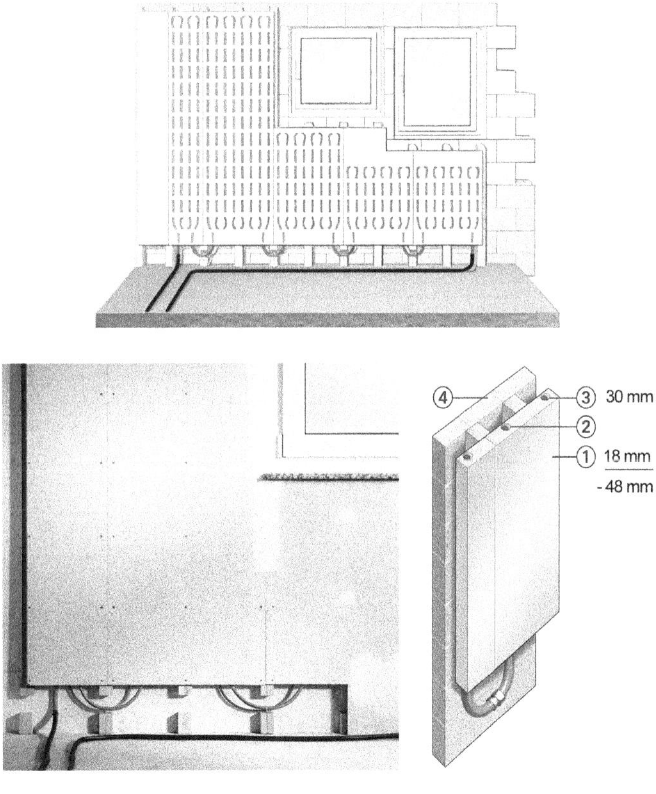

FIGURE 1.11 The construction diagram and connections in the system from [11]: 1 – reinforced gypsum board, 2 – integrated polybutylene pipes, 3 – construction battens, and 4 – wall structure.

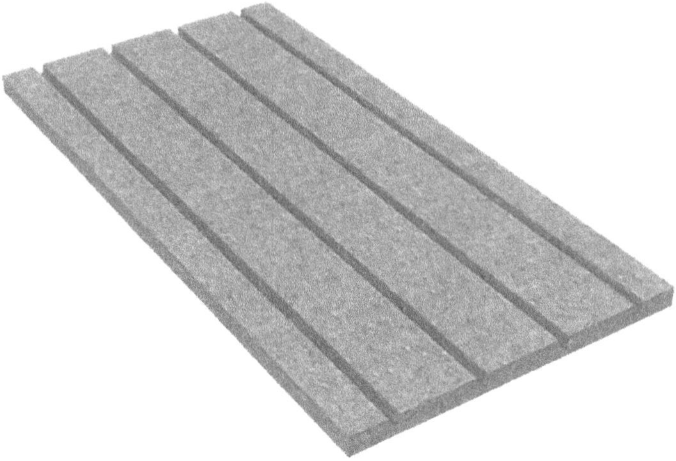

FIGURE 1.12 The panels are based on porous wood fibreboard from [12].

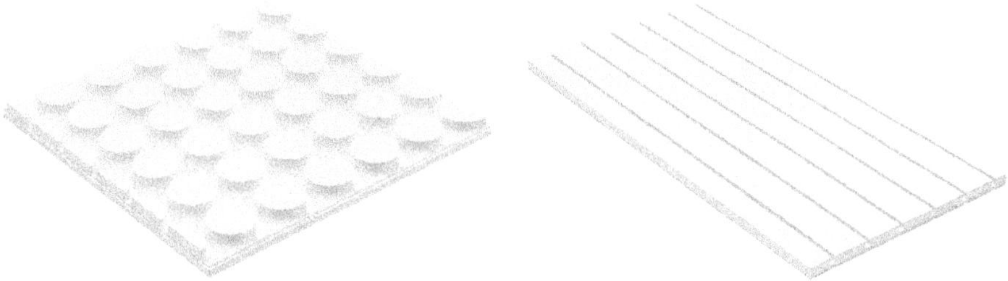

FIGURE 1.13 The panel based on fibre-gypsum boards from [12].

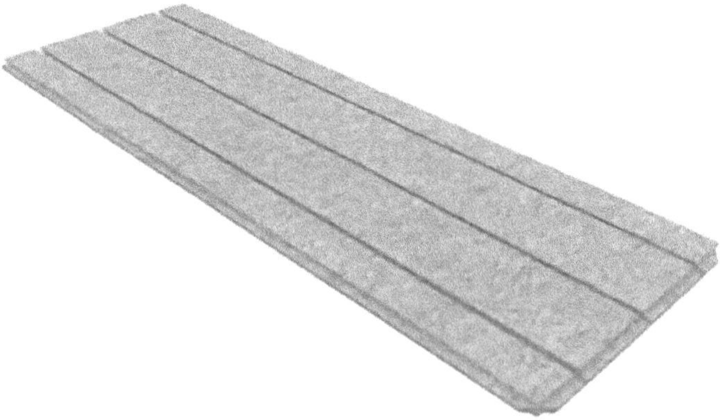

FIGURE 1.14 The panel based on multi-function panels (MFP) boards from [12].

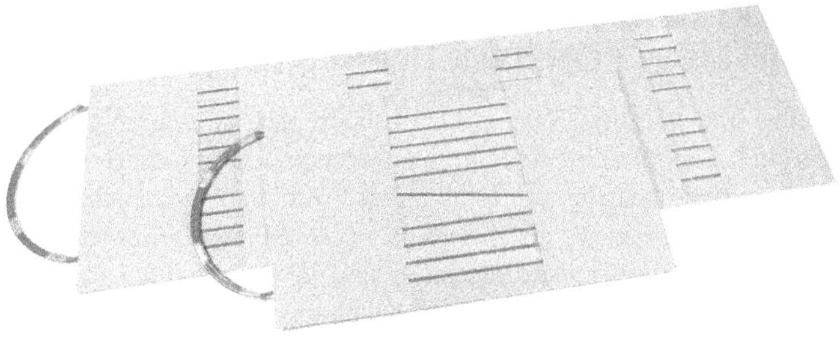

FIGURE 1.15 The fibre-reinforced gypsum board with heating pipes inside [7].

FIGURE 1.16 XPS insulation board covered with aluminium lamella [13].

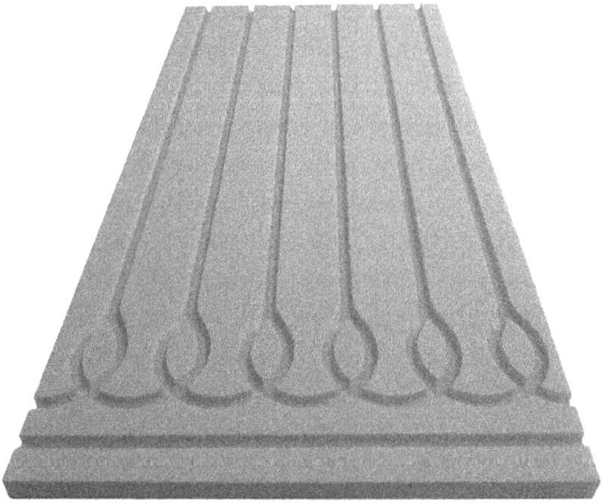

FIGURE 1.17 XPS insulation board without heat dissipation layer (any lamella) [13].

1.3 SUMMARY OF WATER RADIANT HEATING AND COOLING FOR DIFFERENT CONSTRUCTION TYPES

Radiant heating and cooling systems are divided into two main types: "heavy" and "lightweight". The primary structural difference between these two systems is that in "heavy" surface heating, a screed, either "wet" or "dry", is used, whereas in "lightweight" surface heating, no screed is used. Many figures in this chapter describe the most commonly used materials and layers of which these two types of heating and cooling are composed. References to standards for calculating heating and cooling outputs, as well as other parameters characterising surface heating, are also provided. The "heavy" type of surface heating is referenced in numerous national standards, including European standards. It is still widely used, whereas the "lightweight" type is referenced in only one national standard, the Norwegian one [8]. Furthermore, the Norwegian standard [8] applies only to calculating thermal output and flow temperature when the heated floor layer is equipped with a heat-dissipating metal lamella layer. In other lightweight systems, including those without a lamella layer, the thermal efficiency should be determined experimentally, which does not facilitate calculation. To date, relatively few scientists have researched and analysed "lightweight" surface heating and cooling systems, but there is hope that this trend will change in the future. A European standard could then be developed, based on which universally applicable efficiency rules for lightweight surface heating and cooling systems could be designed, both with and without metal heat diffusers. This is possible due to the numerous advantages of this system resulting from the lightweight, heated, or cooled structure described in this chapter. The lightweight construction of the floor, wall, or ceiling layers on which the lightweight heating and cooling system is installed allows for its universal application. The absence of a "lightweight" screed layer can be an advantage if the investor wishes to renovate or modernise the existing building without removing all or part of the existing building partitions. Installing a lightweight surface heating and cooling system on an existing unheated floor or wall, without the need to remove it, is one of the many advantages of this system. This system can also be successfully installed on lightweight wooden ceiling structures. Other advantages that may be worth investigating by researchers include reduced thermal inertia and easier temperature control under changing thermal and humidity conditions within the building, due to the absence of screeds. The need for a quick response to changes in interior temperature may arise when large glazed areas of the building walls, usually located on the south side, can cause a significant increase in temperature when exposed to substantial sunlight. Such solutions (using extensive south-facing glazing) are necessary as one of many requirements in the design of passive buildings, which are increasingly being implemented. In such cases, control in a "lightweight" system will allow for rapid correction of excess heat due to the lack of inertia in the building envelope. Ideally, the lightweight radiant heating/cooling system (LRH/CS) should be connected to a reversible heat pump, allowing for heating in winter and cooling in summer. Some of the advantages of the LRH/CS, mentioned in this chapter, resulting from the floor and wall layers used in this system, will be discussed in more detail in subsequent chapters.

ABBREVIATIONS

kJ	kilojoule
LRH/CS	lightweight radiant heating/cooling system
MFPs	multifunction panels

REFERENCES

1. Adamczyk J., Air heating of rooms. *Construction and Architecture*, 12 (4) (2013), pp. 135–143. (in Polish).
2. EN 1264-2: 2008+A1:2012, IDT. Water-based surface-embedded heating and cooling systems. Part 2: floor heating: prove methods for the determination of the thermal output using calculation and test methods. European Committee for Standardization (CEN): Brussels, Belgium.
3. EN 15377-1:2008, ITD. Heating systems in buildings: design of embedded water-based surface heating and cooling systems. Part 1: determination of the design heating and cooling capacity. European Committee for Standardization (CEN): Brussels, Belgium.
4. Kowalczyk A., *Dimensioning of underfloor heating according to EN-1264.* Materials for lectures and exercises. Faculty of Environmental Engineering. Air Conditioning and Heating Department. Warsaw University of Technology. v. 08–2004–2008, 2004, Warsaw, Poland. (in Polish).
5. Strzeszewski M., *Water underfloor heating.* Materials for lectures and exercises, Institute of Heating and Ventilation, Warsaw University of Technology, v. 0-92–2006, 2006, Warsaw, Poland. (in Polish).
6. Żukowski M., *Underfloor heating.* Handbook for students of Environmental Engineering. Białystok University of Technology Publishing House, 2009, Bialystok, Poland (in Polish).
7. Technical information 864621. Surface heating and cooling. Rehau Unlimited Polymer Solution, February 2017, Poland. (in Polish).
8. NORDTEST Method NT VVS127: UDC 697.1: Approved 2001–11. Floor heating systems. Design and type testing of waterborne heat systems for lightweight structures. Finland.
9. Karpiesiuk J., Chyży T., *Results of experimental tests on the temperature distribution of surface radiators of lightweight construction without screeds.* Rynek Instalacyjny, November 2017, No. 11/2017, year XXV, pp. 45–50. (in Polish).
10. Technical guide part 1 Upnor. Surface heating/cooling, 2012, Poland. (in Polish).
11. Catalog of Fontera System übersicht. Type Fonterra Side 12. Anwendungstechnik Fonterra, 2013.
12. Eco Thermo Floor technical catalogue, 2011, Poland. (in Polish).
13. Elektra Kardo technical catalogue, January 2023, Poland.

2 Electric Radiant Heating Structures

Electric radiant heating can be divided into underfloor, wall, and ceiling heating based on its location. Electric floor heating systems are available as heating cables, heating mats (cables attached to a mounting mesh), or heating foils. Heating foils are most often made of conductive graphite embedded in polyethylene terephthalate (PET) foil or of thin conductive metal alloys (electric wires) embedded in aluminium foil. Like water heating, electric surface heating can also be divided into "heavy" and "light" types. "Heavy" heating occurs when heating cables are installed in or on various types of heavy screeds. "Light" electric heating is considered a floor structure without any screed; for example, heating mats or foils laid on thermally insulating building boards and directly covered with various types of flooring, either glued or laid "dry". Designing electric surface heating is simpler than its water counterpart because there are no energy losses during its transmission at the heating point. It can be assumed that the power of electric heaters equals the design heat load, especially in "light" systems where there is no heat accumulation in the screed.

The heating element in radiant electric heating is resistance wires, made of various types of metal alloys or heating foils with graphite strips as conductors of electrical energy. There are several types of electric heating. The most popular are heating cables, mats, and foils. The construction example of a heating cable from the Elektra catalogue [1] is shown in Figure 2.1.

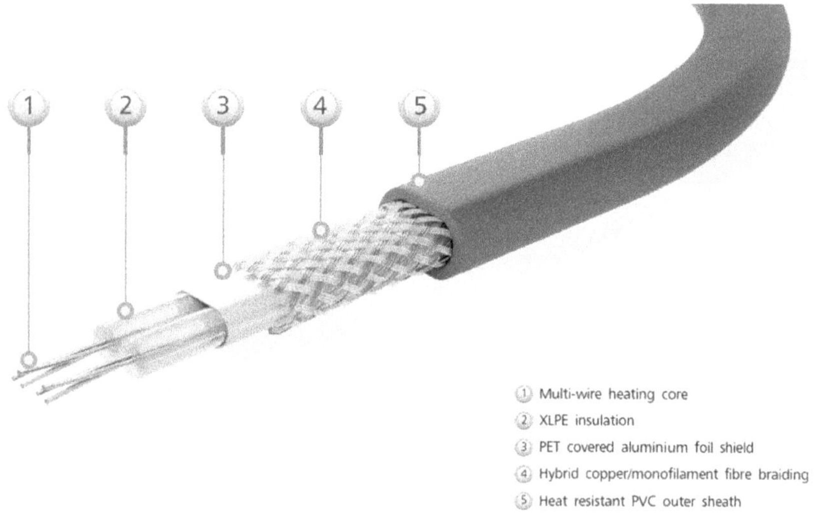

FIGURE 2.1 VCD heating cable design.

DOI: 10.1201/9781003679004-2

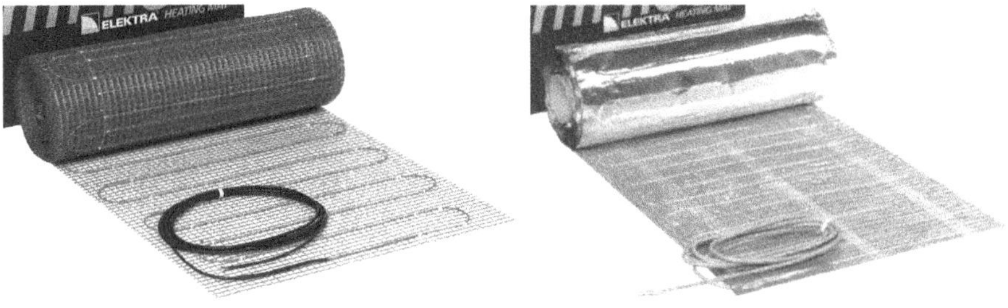

(a) (b)

FIGURE 2.2 The heating mat (a) on a net without foil and (b) on a net with aluminium foil.

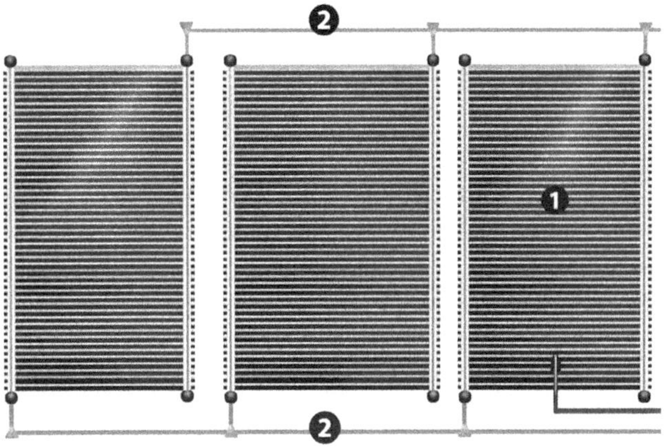

FIGURE 2.3 Heating foil installation diagram: (1) heating foil and (2) heating foil power supply.

The heating mat, drawings of which can be found in [1], consists of a thin wire laid on a fibreglass mesh, without aluminium foil, or with aluminium foil (as in Figure 2.2.)

The heating foil, as shown in the installation manual [2], consists of many graphite strips applied by printing onto a PET foil, which serves as the insulating and supporting layer. On the edges (through strips containing silver compounds), copper strips are connected to graphite strips, forming a heating element. The whole thing is embedded in Laminex foil, ensuring tightness and moisture resistance, as shown in Figure 2.3.

The cables, mats, and heating foils from the Elektra catalogue [1] shown in Figures 2.4–2.11 can be laid on the floor, walls, and ceiling, just like water heating systems. The choice of the type of electric heating depends on the arrangement of layers of the building partition on which the heating system will be installed.

Heating cables include:

- cables installed in the so-called layer wet screed on the floor, using a mesh (Figure 2.4) or mounting tape (Figure 2.5);

FIGURE 2.4 Heating cable in a "wet" screed, mounted on a metal mesh.

FIGURE 2.5 Heating cable in a "wet" screed, mounted on a metal tape.

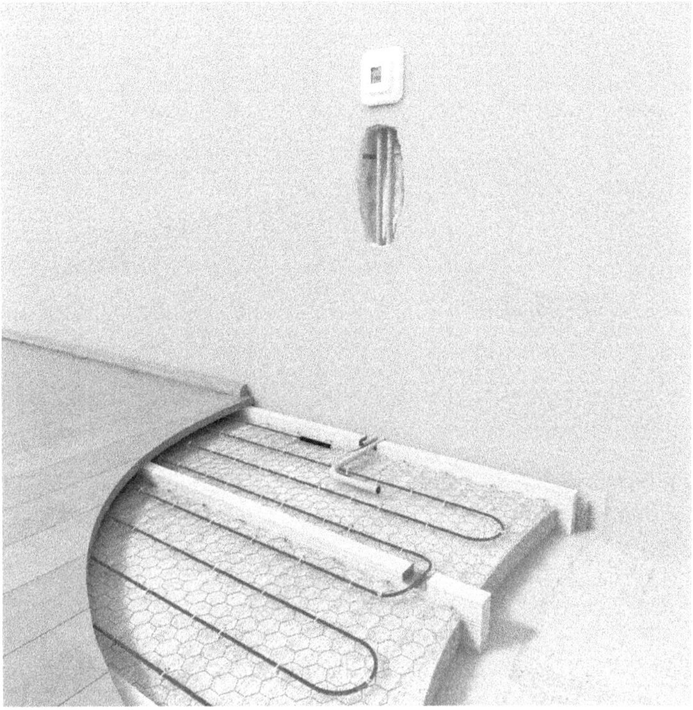

FIGURE 2.6 Heating cable under a wooden floor mounted on a wire mesh.

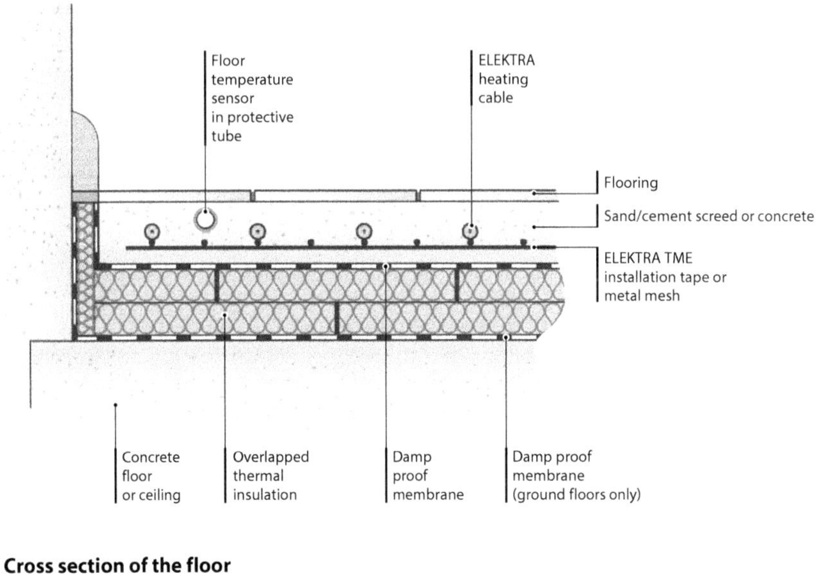

FIGURE 2.7 Cross-section of a storage heating installation with application heating cables.

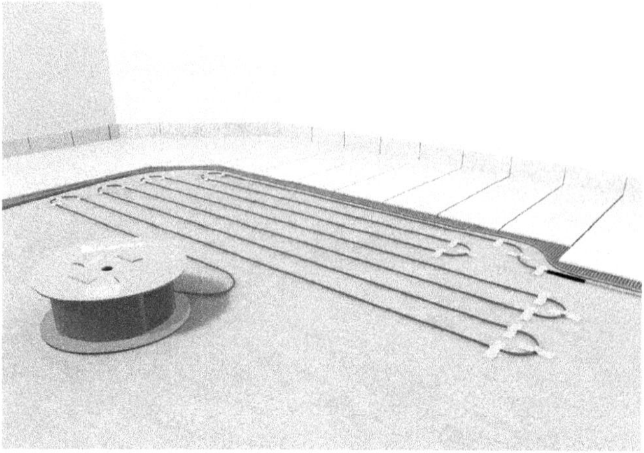

FIGURE 2.8 Thin heating cable pre-attached with self-adhesive tape covered adhesive mortar under the terracotta.

FIGURE 2.9 The heating cable is attached to the wall with hot glue under wet plaster.

- cables mounted under the wooden floor on joists (Figure 2.6);
- cables mounted on the floor in a thick layer of wet screed, constituting an accumulation layer (Figure 2.7);
- thin heating cables mounted in an adhesive mortar (Figure 2.8);
- cables mounted on the wall under a layer of wet plaster (Figure 2.9);
- cables mounted on the wall or floor in the grooves of the insulating layer (Figure 2.10).

When using heating mats, they are installed in a layer of adhesive or self-levelling mortar (after mechanically fastening the mat), most often under ceramic-stone

(a) (b)

FIGURE 2.10 The heating cable is mounted in specially profiled grooves of the board insulating extruded polystyrene, with or without a mesh-adhesive coating: (a) on the wall and (b) on the floor. (Photo from the archives of Elektra Kardo.)

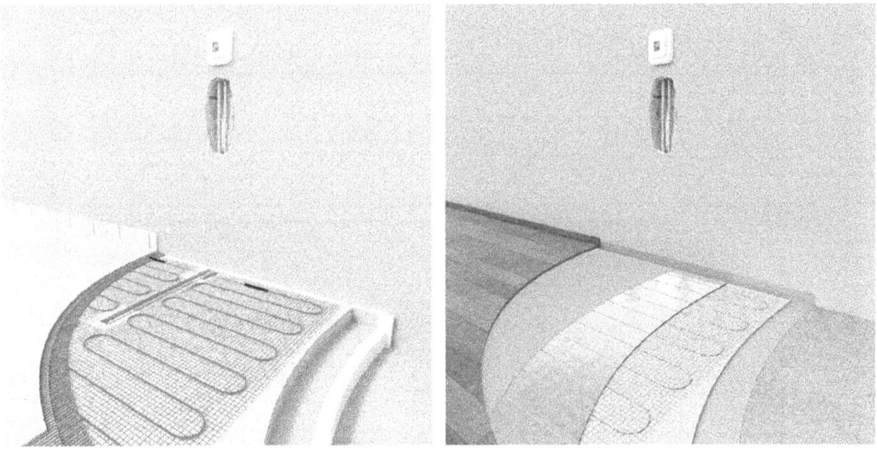

(a) (b)

FIGURE 2.11 Heating mat installation diagram: (a) fixing with adhesive mortar and (b) dry installation, without fixing.

floors or without fastening under panel floors not glued to the structural subfloor (Figure 2.11).

The so-called heavy electric underfloor heating system involves laying cables in a wet screed or heating mats on this screed. Lightweight electric underfloor heating systems include those placed directly on thermal insulation made from insulating building boards, e.g., the Thermopanel system (Figure 2.12):

- the heating mats system and thin heating cables laid in a layer of adhesive or self-levelling mortar directly above the insulating boards, under the flooring,

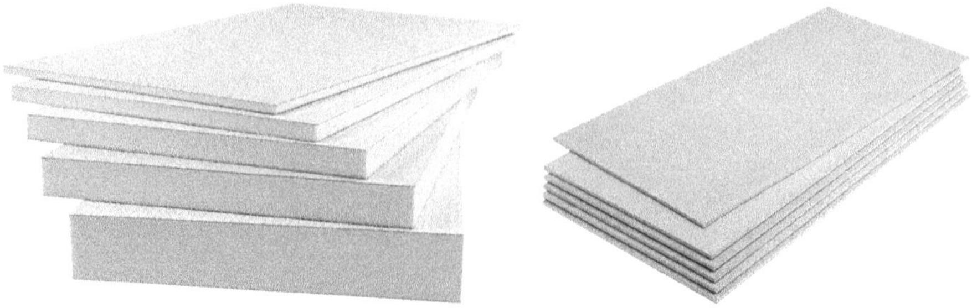

FIGURE 2.12 Building board based on XPS thermal insulation, with an adhesive and mesh layer. (Photo from Elektra Kardo.)

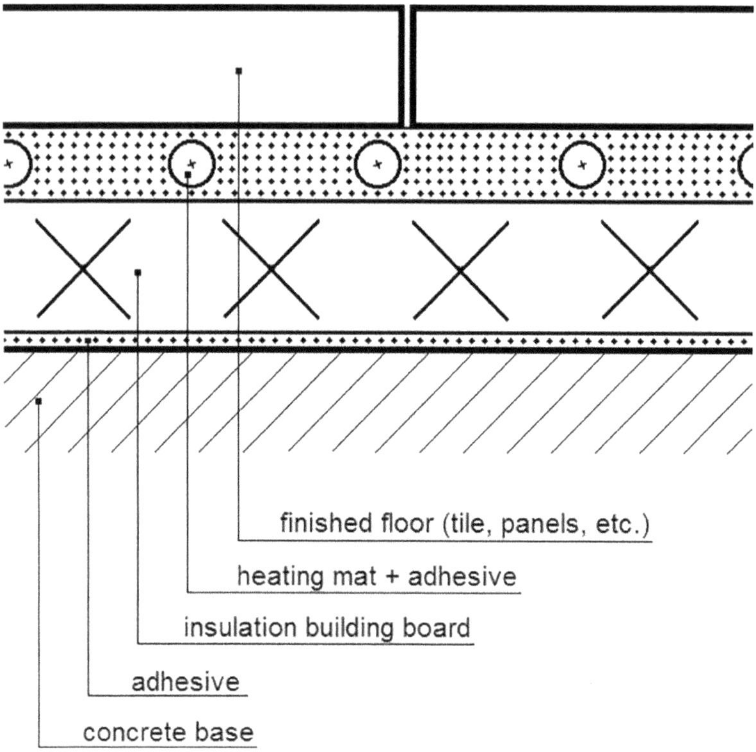

FIGURE 2.13 Cross-sectional diagram of a lightweight floor with an electric heating mat.

- dry-installed heating foils under wall, floor, or ceiling coverings,
- standard (not thin) heating cables are placed in grooves of thermal insulation, on which tiles, terracotta, or other finishing, wall, and floor layers are placed directly (Figure 2.10).

A vertical cross-section example of lightweight electric heating from the Elektra Kardo catalogue [3] is shown in Figure 2.13.

Designing electric radiant heating systems is much simpler than their water equivalent. There are no problems related to the flow of water at different pressures, connection to heat-generating devices of different efficiency and thermal performance, and related standards, nor many complicated designs. In electric heating, the efficiency can be assumed to be 100%, because the source of heat is electricity (powering the heating element at its place of operation). Both types (water and electric) share a similar operating principle, based on better heat dissipation through radiation rather than convection, and the need to control them through thermostat systems, whose purpose is to ensure comfort and reduce operating costs. The selection of devices depends on the design thermal load of buildings, calculated according to PN EN 12831-June 2006 [4] or Strzeszewski and Wereszczyński [5]. Knowing the total design heat load and assuming 100% efficiency of the electric underfloor heating system, one should select heating cables with a power corresponding to the design heat load of the building, taking into account a safety factor of 5%–30%. Then, knowing the type of flooring and the unit power cable, we calculate the spacing between cables in the room in an area free from buildings according to formula (2.1) of Vademecum Elektra [1]:

$$a-a = S/\left(L+0.5P\right) \tag{2.1}$$

Where
- a–a – spacing between wires (cables)
- S – floor surface area on which the heating cable will be laid
- L – length of the heating cable
- P – circumference of the floor on which the cable will be laid

Permanent floor coverings without legs (wardrobes, furniture, carpets, bathtub, shower tray, etc.) are not taken into account when calculating the surface area S. Then, we proceed to install the heating and control system. The heating cables are connected to the thermostats using the so-called cold cables. The thermostat is powered by the electrical switchboard, which should be equipped with appropriate devices to protect against electric shock. Heating mats operate with a constant power per square metre, so all you need to do is match them to the design heat load and install them. However, it is worth remembering that the type of cable or heating mat used, as well as its power, depends on the type of floor covering. Guidelines can be found, for example, in Elektra's Vademecum [1] or Żukowski's manual [6].

2.1 SUMMARY OF THE DIFFERENT TYPES OF ELECTRIC RADIANT HEATING CONSTRUCTIONS

Electric floor or wall heating is one type of many radiant heating systems. Similar to water heating, it can be either a "heavy" type, with a "wet" or "dry" screed in the heated floor layer, or a "lightweight" type, which does not use a screed. The most important types and designs of electric floor heating are listed in this chapter and presented in photos and drawings. Basic principles for the design and installation of this heating system are also provided. Detailed principles for the selection, design, installation, and control of electric surface heating are not provided, as they are fully described in manufacturer catalogues or the literature selected in this chapter. It should also be noted that an electric surface system can only be used for space heating. It cannot be used for cooling buildings.

It's worth mentioning, however, that constant resistance or self-regulating heating cables used for heating buildings can be used for other purposes not described in this chapter. They are ideal for preventing snow and ice from forming on the exterior surfaces of stairs, sidewalks, and driveways. They can also be installed in gutters and downspouts to ensure unobstructed passage during winter or used to protect various sanitary systems, such as water pipes, sewer pipes, storm drains, and other infrastructure.

ABBREVIATIONS

PET polyethylene terephthalate
XPS extruded polystyrene

REFERENCES

1. Elektra Vademecum, Poland edition 09/2021.
2. Heat Decor heating foil installation instructions, Poland, 2017. (in Polish).
3. Elektra Kardo technical catalogue, Poland, January 2023.
4. EN 12831-1:2017, ITD. Energy performance of buildings: method for calculation of the design heat load. Part 1: space heating load, Module M3-3. European Committee for Standardization (CEN): Brussels, Belgium.
5. Strzeszewski M., Wereszczyński P., *Method of calculating the thermal load of buildings according to the PN-EN 12831 standard.* 2nd extended edition, Elektra, Warsaw, 2016. (in Polish).
6. Żukowski M., *Underfloor heating.* Handbook for students of Environmental Engineering. Białystok University of Technology Publishing House, Bialystok, Poland, 2009. (in Polish).

3 General Characteristics of Radiant Heating and Cooling

Radiant heating installed on the floor is typically achieved using coils covered with a wet cement screed. Lightweight surface heating systems are not yet as popular. Numerous studies show that radiant heat transfer, which is over 50%, improves comfort and reduces heating costs when using lower supply temperatures. The optimal floor temperature is around 26°C for heating and no less than 18°C for cooling. The low surface thermal inertia, due to the absence of screeds, allows better control of the heating system and higher heat power. Surface heating systems are most effective on the floor, while cooling systems are most effective on the ceiling. Both systems are more effective when combined with low-temperature heat sources, without any screed. Metal lamellas provide an effective means of distributing heat or cooling. The lack of lamellae can be compensated for by denser coil spacing. Numerous studies, calculations, and analyses indicate that lightweight surface heating and cooling systems are a future-proof solution for energy-efficient buildings due to their short response time to temperature changes, higher thermal efficiency than "heavy" systems, and more effective integration with renewable energy sources thanks to their lower supply temperature compared to traditional surface systems.

Underfloor heating has been known for many thousands of years [1]. Among the mentioned types, systems, and design solutions for surface radiators, the most popular and widely used is underfloor heating with a wet cement screed, i.e., type A according to the PN-EN 1264-2:2008+A1:2012 [2] standard. This choice is influenced by many factors, including installers' opinions on the high thermal comfort of such a radiator, the lack of knowledge of other similar solutions, multiple design studies, and the availability for purchase and the simple, relatively inexpensive installation.

The basic task of every heating and cooling system in a building is to maintain a constant indoor air temperature, providing people with a sense of thermal comfort. To ensure this, the temperature distribution in rooms and on building partitions should be uniform. With an increase in temperature on the surface of the heated floor, as well as on the walls and ceiling (as described by Dzierzgowski et al. [3] and Hao et al. [4]), the air temperature in the room can be reduced by approximately 2 K compared to the traditional heating system, which will not negatively affect the sense of thermal comfort and will also bring measurable economic benefits. The relatively low air temperature is compensated for by the thermal radiation of walls, floors, and ceilings, ensuring thermal comfort, as shown in Figure 3.1 of [5].

Many years of research have shown that low-intensity thermal energy emitted from a large floor surface by radiation and convection results in even heating of the

DOI: 10.1201/9781003679004-3

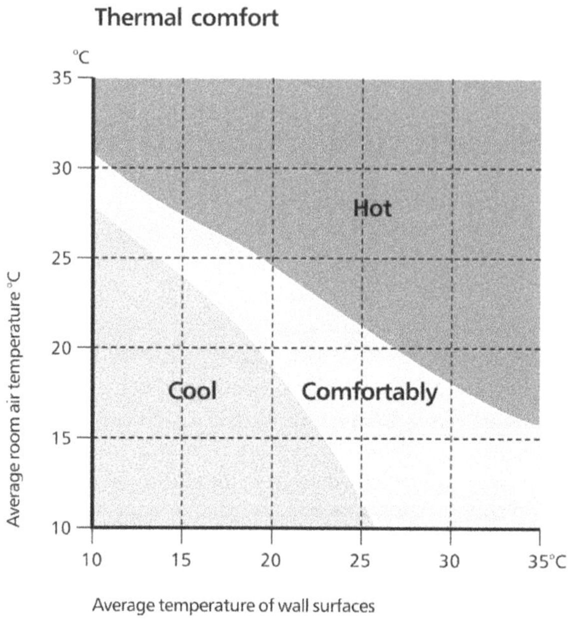

FIGURE 3.1 Human thermal sensation depends upon the average temperature of the air and the average wall surface temperature.

walls and ceiling, thereby making the temperature distribution in the room close to ideal, as shown in Figure 3.2 of the Elektra catalogue [5]. However, one must remember not to exceed the permissible floor temperature specified by the PN-EN 1264-2: 2008+A 1:2012 [2] standard of 29°C, assuming a difference in the supply and return temperature of the heating medium $\sigma = 0$ K in the main (central) heating zone and 35°C in the edge zone (e.g., with extensive glazing of the building), when $\sigma = 0$ K, due to the consequences of producing discomfort to humans. The values of minimum floor surface temperatures for cooling (following the PN-EN 15377-1:2008 standard [6]) are 20°C for sedentary work or rest and 18°C for high activity. With respect to the heating wall only, the maximum temperature by this standard was assumed to be 40°C, and the minimum cooling temperature should be selected so that it does not fall below the dew point temperature.

Based on many studies of underfloor heating installations, the optimal, comfortable floor temperature is 26°C. This is confirmed by Dzierzgowski et al. [3]. The proper location of the heating or cooling medium significantly improves the system's efficiency and operation. By arranging coils, air ducts, cables, or heating foils directly under floors or deeper in layers of cement mortar, we determine the rate of heating or cooling. A material that transmits heat or cold, placed in a building partition with low thermal capacity, will enable quick adjustment of the set temperature in changing conditions inside and outside the building. Variable conditions include the number of people and the operation of household appliances, as well as external conditions, which may be influenced by, for example, sunlight, wind direction and strength, and other factors. When we design floors with concrete layers and standard

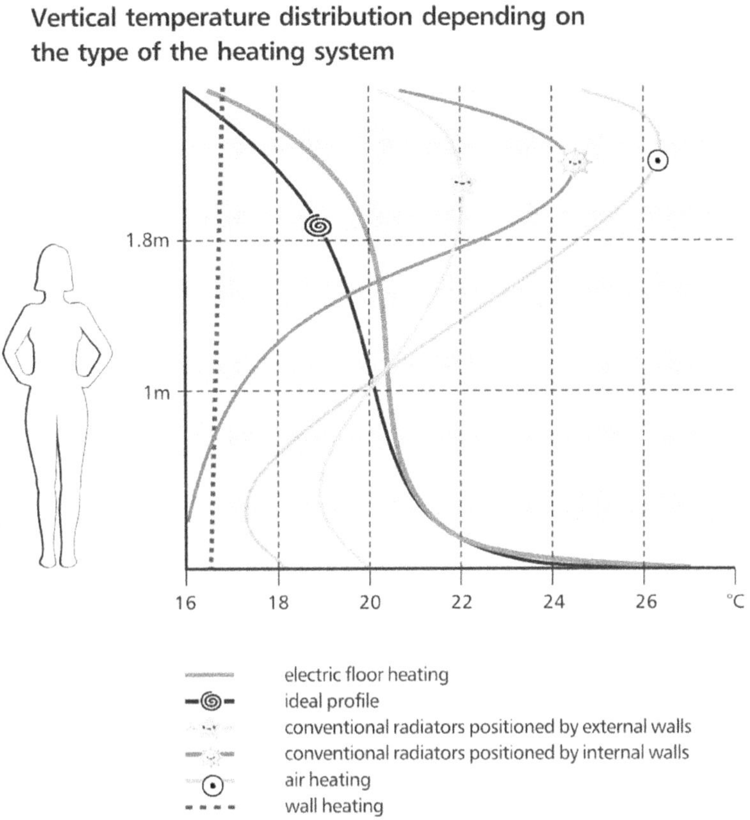

FIGURE 3.2 Vertical temperature distribution depends on the heating system.

thicknesses of 5–6 cm, or the so-called storage layers with a thickness of 8–10 cm, heating or cooling devices can't respond quickly to sudden changes in weather conditions outside or inside the room. One advantage of high-thermal-capacity floors is that they retain heat for a more-extended period in emergencies, i.e., in the event of a prolonged loss of energy supply. Another advantage is lower operating costs, due to much cheaper electricity in the so-called second (night) tariff when we design electric storage heating. In passive houses, where the annual heating energy demand cannot exceed 15 kWh/(m^2/year), additional heating is required, as defined in the publication by Sankowski [7]. It reads as follows: "A passive building is a facility with very low heat demand for heating, not exceeding 15 kWh/(m^2 per year)". Thermal comfort is provided mainly by passive heat sources, such as solar radiation, electrical appliances, and heat generated by the inhabitants. The remaining heat demand during periods of lower temperatures is satisfied by heating the ventilation air or by a low-power heating system. This is due to the possibility of short-term external weather conditions deviating from the assumed (designed) conditions, or an individual's need to create additional comfort with a warm floor or wall. The lightweight radiant system fits perfectly into the conditions and requirements of passive houses, which are to be widely built soon.

It is worth remembering that, regardless of the radiant heating or cooling used with small or large heat capacity, an important element is properly designed control, as described by other authors [8,9]. This aspect is, among others, raised in European legislation, specifically in the ErP (Energy-related Products) Directive regarding climate protection. By 2020, compared to 1990, the European Union (EU) had achieved three aims, based on the 20/20/20 regulation. This regulation assumes, first, a 20% reduction in primary energy use; second, a 20% increase in renewable energy use; and third, a 20% reduction in CO_2 emissions [10]. Between 1990 and 2020, the EU achieved a 34.4% reduction in CO_2 emissions. This means that the EU has achieved its 2020 reduction target. At the same time, Poland recorded a 21% reduction in emissions. Compared to 2010, the energy intensity of GDP in 2020, according to the Central Statistical Office (CSO), decreased by 24.7% (primary) and 20.3% (final); after considering the climate correction, the improvement rate was slightly lower (21.1% and 14.5%, respectively). According to the CSO, the share of energy from renewable sources in the total primary energy supply increased from 19.74% in 2019 to 21.60% in 2020. It can therefore be concluded that the EU assumptions under the 20/20/20 regulation have been met.

As of 1 January 2018, Commission Regulation (EU) 2015/1188 [11] entered into force regarding heating devices using gaseous fuels, liquid fuels, and electricity for domestic use with thermal power of up to 50 kW, and for commercial applications with a product thermal power of up to 120 kW, including their control, taking into account technical, ecological, and economic aspects. The main purpose of the promulgation of this regulation is to reduce the consumption of thermal energy. Therefore, heating devices must be equipped with thermostats that

- Will turn off the heating system when the temperature in the room suddenly drops, which may indicate an open window or door.
- Have a self-adaptation function, i.e., remembering temperature settings and switching on heating devices earlier to fulfil the entered requirements at a given time.
- Can be remotely controlled to change heating or cooling parameters when you are not in the building.

Environmental protection and the effects of global warming have become important factors in evaluating the optimal use of energy, which affects the lives of so many people. About 40% of global energy consumption is due to all types of buildings [12]. The main reason is rising energy consumption for heating, ventilation, and cooling. This causes increased pollution of the natural environment, resulting, for example, in high levels of suspended dust (such as PM2.5 and PM10) from the combustion of fossil fuels [13]. This phenomenon, called "London-type smog", is more intense in winter, especially in dense urban areas, where heating devices, such as boilers, are most often fired with fuel of questionable quality. In addition to its negative impact on human health, it also accelerates the corrosion of equipment and buildings. To reduce the effects of fuel combustion, intensive research should be performed to reduce such energy consumption by devising and applying new technologies and material solutions, as well as

- Ensuring improvement of the thermal insulation of buildings in accordance with the regulation of the Minister of Transport, Construction and Maritime Economy of 2013 [14], thus reducing the values of the primary energy ratio index for non-renewable primary energy.
- Introducing advanced exhaust gas treatment methods.
- Reducing net emissions by at least 55% (compared to 1990) by 2030, through the introduction by the European Commission on 14 July 2021, of the "Fit for 55" package of legislative proposals under the European Green Deal, which aims to strengthen the EU's position as a global climate leader. The package aims to modernise existing legislation in line with the EU's 2030 climate target, helping deliver the transformative changes needed across the economy, society, and industry to achieve climate neutrality by 2050.

The proposed regulatory framework covers, among other things, the energy sector (including heating and cogeneration systems), land use and forest management, the road transport sector, and energy taxation. All EU legislative activities are aimed at environmental protection, with restrictions on the use of primary energy sources such as gas, oil, and coal, and their replacement, to the greatest extent possible, with machinery, devices, and technologies powered by renewable energy. It is all the more important to look for materials and technologies that reduce energy demand, including, among others, lightweight radiant heating and cooling systems (LRH/CS).

3.1 EFFICIENCY OF "HEAVY" WATER HEATING AND COOLING SYSTEMS

Given the importance of mitigating the effects of global warming, it is vital to minimise waste heat and reduce the combustion products emitted into the atmosphere, such as carbon dioxide, sulphur dioxide, nitrogen oxides, dust, and other pollutants. A significant source of these pollutants is households that require heating during cold outdoor temperatures (the heating season) and cooling in the summer, especially in areas close to the equator. It is necessary to use effective heating and cooling systems that, thanks to improved operating parameters, will reduce the negative impact on the environment. One such parameter is the thermal efficiency of heating devices. The greater the efficiency of heating devices, the higher their thermal power.

The many advantages of radiant heating and cooling systems make them increasingly popular. These include

- high thermal comfort, and even temperature distribution,
- aesthetics of the rooms, due to the concealment of heating systems in the floor, wall, and ceiling,
- improved health and hygiene conditions,
- lower energy consumption, and cooperation with devices using renewable energy sources,
- quiet operation.

Rhee and Kim [15], based on a review of 50 years of literature (1965–2014), provide information on comfort, heat transfer, control, and energy simulations for radiant heating and cooling in buildings. They stated that a radiant heating system is one in which the radiation component exceeds 50% of the total heat expenditure. This allows lowering the air temperature inside the room without compromising thermal comfort. Similarly, when we want to cool a room, cooling surfaces are better at removing heat from people and spaces through radiant heat transfer than conventional air conditioning systems. The slight difference between the set air temperature and the radiant heating/cooling temperature means that such a system can be defined as a heating, low-temperature system [16] or as an air conditioner when the cooling temperature is relatively high [17]. This is important in both cold- and mild-climate countries to reduce energy consumption. Babiak et al. [17] provide the parameters of the heat transfer coefficient between the heating/cooling surface and the air space, which depend on the place where the system is installed. This was confirmed in tests of a floor heater conducted by Olsen [18], which reported a heat transfer coefficient of 11 W/m^2·K; with smaller temperature differences between the floor and air surfaces, it is approximately 9 W/m^2·K. More than half of the total heat transfer coefficient, approximately 5.5 W/m^2·K, results from radiation. The maximum thermal power in the central zones is approximately 100 W/m^2 at an air temperature of 20°C. In the edge zones, it is 165 W/m^2 at approximately 1 m from the wall/window and, on average, 116 W/m^2 at 4 m. The heat transfer coefficient for underfloor cooling is only 7 W/m^2·K. In most buildings, the maximum cooling efficiency of the floor at a room temperature of 20°C is 42 W/m^2. However, when sunlight directly falls on the floor, e.g., with extensive glazing, it may be necessary to increase the cooling efficiency beyond 100 W/m^2.

The type of system and the thermal resistance of the floor, wall, or ceiling cladding also influence the heat flux density. It is higher when the surfaces emitting heat or cold have higher thermal conductivity and lower when the lining resistances are high. An example would be to compare the thermal efficiency of a type B radiator structure with that of a type G radiator, both using aluminium heat spreaders. Thermal efficiency in type B, according to Olsen [18], is approximately 2 times higher than in type G, with coil spacings of 150 and 300 mm, and is mainly influenced by the lining type, with a higher thermal conductivity coefficient in type B. The test results showing the maximum thermal power values for heating and cooling are given in Table 3.1.

TABLE 3.1
Important Heat Transfer Parameters for Heating and Cooling

Partition		Total Heat Transfer Coefficient [W/m^2·K]		Permissible Temperatures on the Surface [°C]		Maximum Thermal Power [W/m^2]	
Type	Zone	Heating	Cooling	Heating Max.	Cooling Min.	Heating	Cooling
Floor	Medium	9–11	7	29	20	99	42
	Edge	9–11	7	35	20	165	42
Wall	x	8	8	40	17	160	72
Ceiling	x	6	9–11	27	17	42	99

Source: Based on [18].

The analysis of using "heavy" underfloor heating as a cooling system was made by Werner-Juszczuk [19]. To do this, she used numerical simulations performed with a 2D model to evaluate the Workbench 19.2 ANSYS Steady-State Thermal Analysis System program, which uses the mechanical APDL solver. The analysis included a type A underfloor heating system, with pipes (16 × 2 mm) attached to thermal insulation and placed in a 6.5 cm thick cement screed. The heat transfer coefficient values were adopted from the ISO 11855 standard [20] with a value of 6.5 $W/m^2 \cdot K$. When the coil spacing is 0.1 m, and the thermal resistance of the floor surface is 0.02 $m^2 \cdot K/W$, then, at a floor surface temperature of 20°C, it will achieve a heat flux density of approximately 40 W/m^2. This value confirms the data from Table 3.1. This thermal power will be achieved by a "heavy" system at an air temperature of 26°C, with the water temperature approximately 16°C (Figures 3.3 and 3.4).

Figures 3.3 and 3.4 show that the temperature depends on the floor surface and heat fluxes, as well as on the temperature inside the coil or the thermal resistance of the floor. These data also rely on the coil spacing and the room temperature.

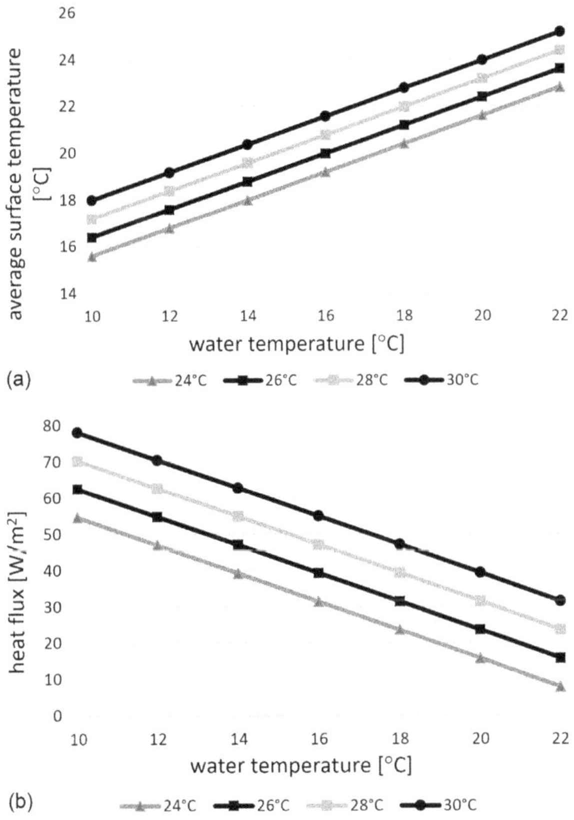

FIGURE 3.3 The relationship between the cooling temperature of the water in the coil, and (a) the average temperature of the floor; (b) heat flux density at different room temperature variants, when the coil spacing is 10 cm and the thermal resistance of the floor is 0.02 $m^2 \cdot K/W$ [19].

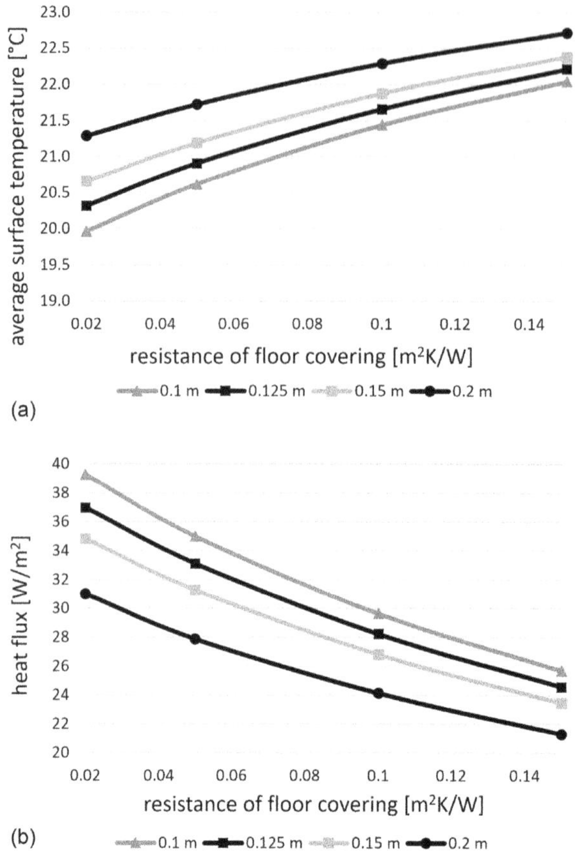

FIGURE 3.4 The relationship between the thermal resistance of the floor, and (a) the average temperature of the floor; (b) heat flux density at different coil spacing variants when the water temperature is 16°C and the room temperature is 26°C [19].

The given graphs can be used to determine the theoretical cooling capacity of "heavy" underfloor heating. It should be borne in mind that the theoretical floor cooling efficiency can be very high, but is limited by the minimum surface temperature of the cooling floor, as established by the standard [21]. Failure to comply with the standard's requirements may lead to water condensation on the floor and reduced user comfort.

According to Feustel and Stetiu [22], using standard air conditioning solutions is not economical because they require heating or cooling air, which has low heat capacity. It involves the use of higher fan power, which increases noise in the room. In this case, there is no uniform transfer of heat or cold, which may lead to people staying near the air conditioner for a long time. Stetiu [23] showed that using a radiation cooling system results in 30% less energy consumption and a 27% reduction in heat demand compared to convection systems. The possibility of using large cooling surfaces with radiant systems means that lower power consumption produces a better effect than in conventional heating and cooling systems.

Radiant cooling systems can be considered an effective technology for reducing indoor temperatures due to their higher user comfort [24,25] compared to conventional air conditioners and their greater energy-saving potential [26,27]. In a similar study, Sastry [28] confirms that energy savings amount to 34% compared to a conventional system. Tantiwichien [29] also proves that energy savings from radiant heating can reach 41% compared to traditional air conditioning. In addition, Yau [30] describes another essential advantage of surface cooling: the absence of direct, local airflow impact on a person, which may contribute to illness. The ASHRAE publications [31] recommend three cooling methods: radiant cooling panels (RCPs), embedded surface cooling systems, and thermally activated building systems, in which heating and cooling pipes are placed in the building structure, e.g., a concrete ceiling. The research results [32] show that surface cooling technology provides better thermal comfort than convection cooling technology while also providing lower energy consumption. Thus, a person in a surface-cooled room feels more comfortable than in a conventional cooling system.

Another way is to mount renewable energy sources (RES) (e.g., geothermal, solar, and wind), which are effective when we can utilise the low temperatures of the energy carriers attached to the devices. Thanks to this, we increase the system's efficiency, benefit from the thermal comfort provided by radiant heaters, and also help address the climate of our planet.

For many years, many countries have been taking advantage of the apparent benefits of radiant heating. In recent years, there has been a noticeable tendency to exploit this system universally for both heating in winter and cooling in summer. According to Olsen [18], in Germany, Austria, and Denmark, 30%–50% (and in Korea over 90%) of new residential buildings by 2001 had underfloor heating installed. According to Tye-Gingras and Gosselin [33], this type of heating is also successfully used in commercial and industrial buildings due to its effective heat exchange and comfort.

Since the floor, in particular, may come into direct contact with the human body, as stated in ANSI/ASHRAE Standard 55:2010, ISO 7730:2005, and ISO 11855-1:2012 [20,21,34], it is recommended to adopt a comfortable temperature between 17°C and 29°C, depending on the season, lifestyle, and the quality of the building's insulation. Song [35] states that a higher floor temperature should be implemented in buildings with poor thermal insulation. Research on the optimisation of the thermal insulation used by Cvetković and Bojić [36] in four heating systems, i.e., floor, ceiling, wall, and the newly implemented floor-ceiling systems, showed that when using surface heating, good thermal insulation is a necessary condition for this system to be efficient. However, where thermal insulation is poor or absent, a system based on heat exchange through convection will be more efficient [37] because the losses that significantly impact comfort through radiation are minor in this system, and increasing radiation does not significantly improve thermal comfort. From a practical point of view, an assessment of the heat transfer coefficient, the temperature of the heating and cooling surfaces, and the air temperature is needed to determine the heat flux density q, following the standards PN-EN 15377-1:2008 [6] and ISO 11855-1:2012 [20].

It is

- in the underfloor **heating** and ceiling **cooling** system:

$$q = 8.92\left(\theta_{S,m} - \theta_i\right)^{1.1} \left[\mathrm{W/m^2}\right]$$

- in the wall **heating and cooling** system

$$q = 8\left(\left|\theta_{S,m} - \theta_i\right|\right)\left[W/m^2\right]$$

- in the floor **cooling** system

$$q = 7\left(\left|\theta_{S,m} - \theta_i\right|\right)\left[W/m^2\right]$$

- in the ceiling **heating** system

$$q = 6\left(\left|\theta_{S,m} - \theta_i\right|\right)\left[W/m^2\right]$$

where

q – heat flux density of the radiator
$\theta_{S,m}$ – average floor surface temperature
θ_i – air temperature

These formulas, according to Kim and Olsen [38], show that the floor with a thermal efficiency of 100 W/m^2 of heating has a cooling efficiency of 40 W/m^2 and vice versa; with an efficiency of 100 W/m^2 cooling, the heating efficiency is 40–50 W/m^2.

3.2 EFFICIENCY OF "LIGHT" WATER HEATING AND COOLING SYSTEMS

In the summary of knowledge from the literature up to 2015 [15] on radiant heating and cooling over the last 50 years, many directions for further action were identified to increase thermal comfort, energy efficiency, and, at the same time, energy savings and the use of renewable sources. All this will also help protect the natural environment. One of the suggestions in the summary [15] was to develop a universal heating and cooling model, using which we will obtain the lowest possible heating temperature and a high cooling temperature with high efficiency, along with a control system that responds quickly to changes in the building's environment. A lightweight surface radiator has many features that meet these criteria. This is indicated by its advantages (compared to the so-called "heavy", traditional surface radiators), described in the literature and standards.

3.2.1 Wall-Mounted "Light" Heating and Cooling Systems

One way to increase the efficiency of heating devices may be by using light wall surface radiators, as described by Karpiesiuk and Chyży [39]. Wall heating/cooling systems are now an alternative or supporting solution for underfloor and ceiling heating/cooling needs. They are installed when the floor system has insufficient thermal power, or in rooms where it is not possible to change the floor structure, for example, in rooms that are being modernised. These systems can be installed in residential, commercial, and industrial buildings. One can combine surface, underfloor, and wall heating. This allows for high thermal power, which is essential in large rooms. Dual heating also works well in small rooms with a large floor area, where greater heating power is needed, e.g., bathrooms. The technical guide [40] recommends wall heating for attics. By installing surface heating under the plasterboard structure, the problem of moisture is eliminated. When installing radiators on internal walls without thermal insulation, adjacent rooms must have similar temperature requirements. In the case of installation on external walls, one must make sure that the structure of these walls complies with the requirements of the standards PN-EN 15377-1:2008, PN-EN 12831-June 2006, PN-EN 15377-2:2008, and PN-EN 15377-3:2007 [6,41–43]. Technical conditions required of buildings under the Regulation of the Minister of Infrastructure of 12 April 2002 detailing the technical conditions that buildings and their locations must meet.

The minimum thermal resistance values for thermal insulation of heating/cooling building partitions are given in Table 3.2 [44].

The advantages of wall radiant heating are included in the guide [45], and according to Karpiesiuk [46], which are given below:

- Close to ideal temperature distribution in the room, which results in high levels of thermal comfort.
- The temperature of the heated wall can be much higher, up to 40°C, compared to underfloor heating (maximum 29°C in the occupied zone), which results in greater thermal power.
- The small thickness of the wall cladding above the heating pipes, and therefore low thermal resistance, results in low thermal inertia, which facilitates temperature control.
- No procedures related to the installation and care of wet screeds are required, which extend the installation time of the cladding layers.
- They are ideal for heating using heat pumps or other energy carriers, with a low temperature of the heating medium in the range of 28°C–35°C.
- For small surfaces, it may be cheaper than using the traditional version with wet plaster installation, while maintaining the same technical parameters.

TABLE 3.2
Minimum Value of Thermal Resistance (m^2 K/W) of Installed Thermal Insulation below the Heating Pipes in the Heating/Cooling System

Adjacent	Heated Room Below or Adjacent	Unheated or Temporarily Heated Room Below, Adjacent or Directly on the Ground[a]	Temperature Below or External Design Temperature $\theta_d \geq 0°C$;	$0°C > \theta_d \geq -5°C$;	$-5°C > \theta_d \geq -15°C$
Thermal Resistance R	0.75	1.25	1.25	1.50	2.0

[a] The value should be increased when the groundwater level is ≤5 m below the supporting layer.

The lightweight surface heating/cooling system described in the Guide [40] is also effective for room cooling because

- Small distances between pipes increase efficient cooling, even with a relatively high inlet temperature of water supplying the system.
- The possibility of reducing the pressure by using short heating/cooling circuits, or increasing the pipe diameter.
- It enables better cold flow, due to low thermal inertia, especially when using wall materials with good thermal conductivity, with conductive diffusers (lamellas).
- Thin-wall panels allow better control over the temperature close to the dew point.

Research on the use of wall radiant heating in various variants of lightweight heating systems at different supply temperatures was described by Koca et al. [12]. The heat transfer coefficient between wall-heating panels and the room air was measured for three types of wall-mounted surface radiators, with the supply temperature ranging from 30°C to 42°C in the research module, in increments of 2°C. It has been shown that thermal radiation can achieve at least 50% heat transfer in a radiant heating system [15]. This allows the use of lower supply temperatures for heating and higher ones for cooling, thereby increasing energy efficiency, reducing energy consumption, and providing thermal comfort at a much lower temperature than with other heating systems [15]. This process allows us to successfully utilise the latest devices, such as heat pumps, solar collectors, and photovoltaics. In this way, we recover energy from renewable, geothermal, ground sources, waste heat recovery, solar energy, etc. The research described by Koca et al. [12] was performed in a special research chamber, with a lightweight radiant heating system (LRHS) installed on the western and northern walls, as shown in Figure 3.5.

FIGURE 3.5 Wall-mounted surface radiator.

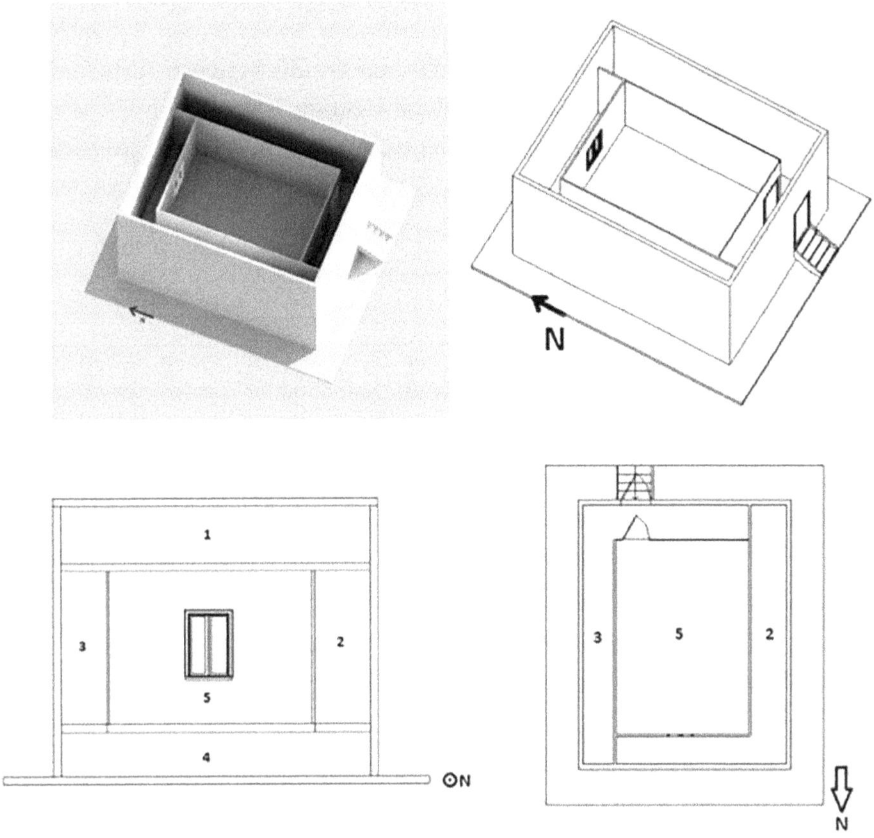

FIGURE 3.6 Research chamber.

The chamber had a volume of 24 m^3 and a height of 3 m (Figure 3.6). It was designed to yield realistic test results under various climatic conditions and, if possible, realistic living conditions by installing a door and a window.

In the cited literature, it is recommended to adopt a constant value of the radiation heat transfer coefficient U_r equal to 5.5 W/m^2·K, which is justified by many research results on this coefficient, while the convective exchange coefficient U_c may vary within the range of 0.3–6.5 W/m^2·K, depending on the location of the heated/cooled surface [12]. The ISO standard [20] and the European standard [47] provide the values of the heat transfer coefficient U_{total}, depending on the position of radiant heating/cooling, which are as follows:

- 10.8 with underfloor heating and ceiling cooling
- 8 for wall heating and cooling
- 6.5 with underfloor cooling and ceiling heating

Koca et al. [12] report that the average U_{total} value for the heated wall, depending on the variants of heating panels and the water supply temperature, was 8.72 W/m^2·K,

TABLE 3.3
Comparison of Heat Transfer Coefficient Results between Tests for a Different Type of Lightweight Radiant Heating System

Literature	Type of Radiator	U_c [W/m²·K]	U_r [W/m²·K]	U_{total} [W/m²·K]
[48.] (Causone, 2009)	Ceiling	0.3	5.6	5.8
	Wall	—	—	8
[6.] EN 15377-1	Floor	—	—	10–11.1
	Ceiling	—	—	6
[12.] (Koca et al., 2014)	Wall	2.25–3.22	5.38–6.25	8.16–9.28

and that the average radiation heat transfer rate was about 68% across the three variants. It is worth noting that the arrangement of heating panels on the walls has a significant impact on the convection heat transfer coefficient U_c and a small impact on the radiation heat transfer coefficient U_r. These results are similar to those achieved by other scientists [48,49] and to the standards cited above. On this basis, it can be concluded that the heat transfer coefficients for surface wall heating are lower than those for underfloor heating and higher than those for ceiling heating. This is shown in Table 3.3.

Acikogoz takes a slightly different position [50]. Research performed using theoretical calculations and the Engineering Equation Solver (EES) program on the radiant and total heat transfer coefficients of surface-heated walls on various surfaces of rooms shows that the average radiation coefficient is in the range of 5.4–5.5 W/m²·K. However, the total heat transfer coefficient through heated walls was much higher than stated in standards and other tests, i.e., 10.2–10.8 W/m²·K. The average share of radiation in total heat transfer was 64%–67%. This is thus another study confirming the thesis of the stability of the U_r parameter, adding only the information that it changes slightly, or even decreases, with increasing distance between the heated and unheated walls. Additionally, it has been shown that U_c and, consequently, U_{total} increase temperature differences between the heated and unheated surfaces. However, regarding the size of rooms, U_{total} decreases slightly as the floor area increases.

A comparison of the thermal efficiency, costs, and CO_2 emissions of surface and convector wall radiators was conducted by Bojić et al. [51]. The experiments were performed in a two-story, single-family building with a total area of 301 m² and a wall heat transfer coefficient of 0.57 [W/m²·K]. In the tested house, surface heating was installed only on the external walls of heated rooms, totalling 212 m², with an average water temperature in the heating plates of 37°C. Some wall radiators were installed with additional thermal insulation (RzI), and others without (RbI). Similarly, in each heated room, convection heaters without (KbI) and with additional thermal insulation (KzI) were installed on the external walls. The whole thing was heated by a condensing gas furnace. The calculations were performed using EnergyPlus [52], and the test results are presented in Figure 3.7.

These studies show that, in a house without additional thermal insulation, a heating system with convectors consumes 21% more primary energy than surface

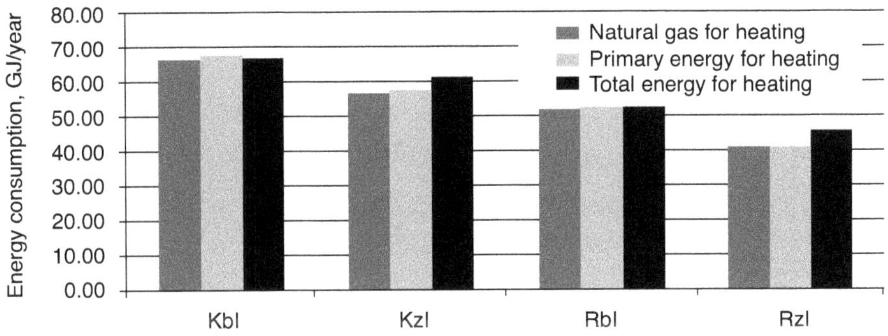

FIGURE 3.7 Energy consumption during the heating season. RzI – wall radiator with additional thermal insulation, RbI – wall radiator without additional thermal insulation, KzI – convection heaters with additional thermal insulation, KbI – convection heaters without additional thermal insulation.

wall-heating systems (RbI), and, with additional thermal insulation, a heating system with convectors (KzI) consumes 28% more primary energy than surface wall heating (RzI). RzI is the most energy-efficient and has the lowest negative impact on the environment. However, this solution is also the most expensive. The wall-heating system is more sensitive to thermal insulation than the convection heating system. This is confirmed by the results of research by Le Dréau and Heiselberg [37], which showed that heating with convectors was less sensitive in buildings with weaker thermal insulation. Acikogoz [50] proves that lightweight radiant wall heating, in which the heat supplied to the radiator is close to the heat emitted, has a thermal efficiency similar to that of floor radiators or even higher, and not, as reported in the literature above, reduced by approximately 17%–23%. Similarly, wall cooling of lightweight radiators, due to their low thermal inertia (i.e., good cold emission), may exhibit higher heat transfer coefficients than those specified in PN-EN 1264–5:2008 [53]. Table 3.4 summarises the test results regarding the heat transfer coefficient for wall heating and radiant cooling. The given data show significant differences in U_{total},

TABLE 3.4
Comparison of Heat Transfer Coefficient Results for Heating/Cooling Radiant Wall, Based on Research and Standards from Selected Literature

Literature	Wall Heating/Cooling	U_c [W/m²·K]	U_r [W/m²·K]	U_{total} [W/m²·K]
Causone et al. [48]	Heating	—	—	8
	Cooling	—	—	—
Koca et al. [12]	Heating	3.15	5.47	8.73
	Cooling	—	—	—
Acikogoz [50]	Heating	3.5	5.5	10.5
	Cooling	—	—	—
EN 1264-5 [53]	Heating	—	—	8
	Cooling	—	—	8

with values over 20% in the range of 8–10.5 [W/m²·K]. From this, it can be concluded that the data used in the standards need not be accurate, and further experiments are still needed to make the results credible.

3.2.2 Underfloor "Light" Heating and Cooling Systems

Thermal properties and thermal comfort of low-height (ultra-thin) surface radiators, also known as lightweight surface radiators, are still not included in European standards. They were, and still are, tested in the most popular position, i.e., on the floor. In 2015, research on this type of radiator was carried out at the Białystok University of Technology as part of engineering work by Paweł Karpiesiuk under the supervision of Prof. Żukowski [54], entitled "Research on the temperature field during surface heating", as well as other experiments [55]. This research used both mathematical and experimental calculations. The mathematical model of the thin radiator [55] (shown in Figure 3.8), which neglects heat losses from the floor, and the research models of the aforementioned works showed excellent thermal stability, mainly due to the high thermal efficiency of radiation.

Research has proven that in this type of radiator, the ability to heat transfer radiation accounts for 60% of the total heating, which is much more than in typical heating systems. In the engineering work [54], the dependence of the density of the heat flux emitted by various models of radiators with dry screeds on the supply temperature was determined, including a thin radiator without screeds (Figure 3.9). This radiant heater, consisting of an adhesive mortar reinforced by the fibre mesh, was placed directly under the tiles. All these floor layers were put on the insulating board. The following conclusions were drawn from the research:

- The most energy-efficient type was the ultra-thin floor model (variant VII – without screed), as shown in Figure 3.10. This model heated up the fastest and delivered heat to the surroundings in the shortest time. The remaining variants involved various configurations with dry screeds.

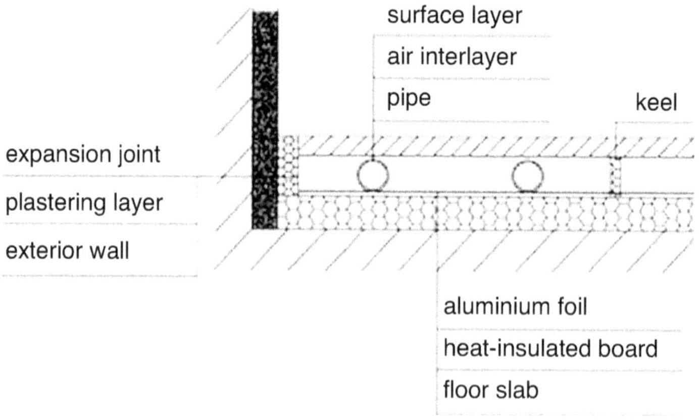

FIGURE 3.8 Research model.

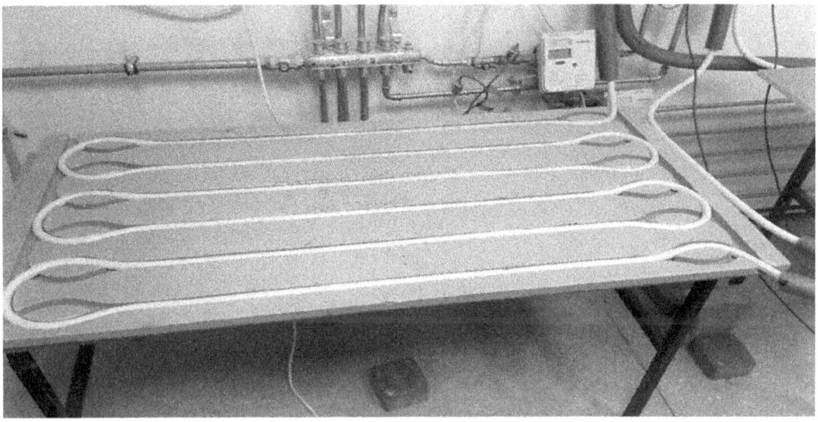

FIGURE 3.9 Research model of an ultra-thin surface radiator on which is superimposed directly the mesh embedded in adhesive and tiles.

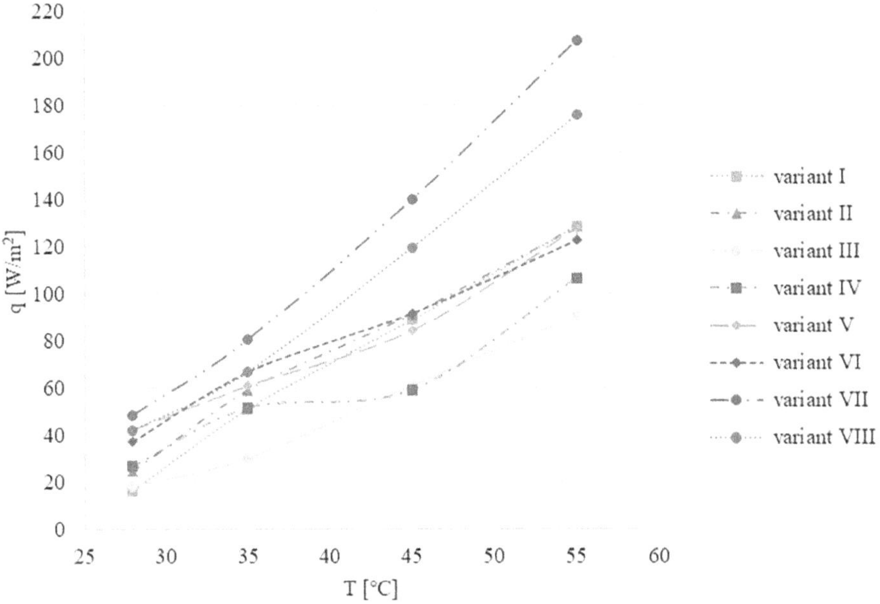

FIGURE 3.10 Summary of the results of eight tests conducted with various types of dry screeds (variants I–VI and VIII) and without screed (variant VII).

- The ultra-thin floor model is effective when powered by RES because, at a low supply temperature of 28°C, the heat flux density q was the highest among comparable models.
- The ultra-thin surface radiator is the cheapest system among those tested, and at the same time, thermally effective; as such, it can be recommended and used by customers in the future.

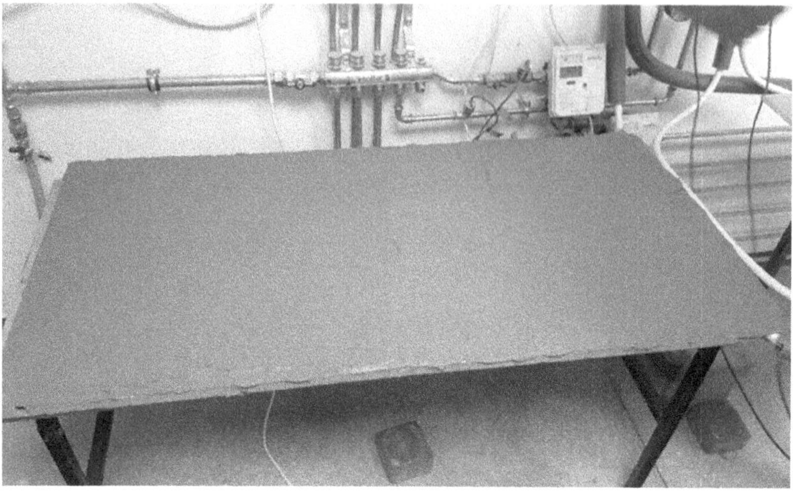

FIGURE 3.11 Model of a lightweight surface radiator without screeds.

Based on the experiment performed in engineering work [54], as shown in Figure 3.11 [56], the temperature distribution on the surface of the radiator without screeds and its course during heating and cooling were determined. The radiator is made of rough 4-cm-thick XPS (extruded polystyrene), without heat dissipation elements. The heating coil from a multilayer pipe, with a diameter of 16×2mm, type PE-RT/AL/PE-RT, was placed in milled grooves, using a spacing of 10cm. The entire structure was covered with a C2S1 adhesive mesh layer and tiles.

The dependence of the heat flux density q emitted from the surface of a lightweight radiator on the supply temperature is presented in Figure 3.12, where the lower, dashed horizontal line indicates the limit value of the flux in the central zone (calculated for a temperature of 29°C), and the upper one in the edge zone (for a temperature of 35°C). These values should not be exceeded due to negative effects on human physiology. Hence, the maximum supply temperature at the limit flux value is only 37°C (Figure 3.12).

The article by Żukowski and Karpiesiuk [56] lists many of the advantages of a lightweight radiator, such as low thermal inertia, low weight, low thickness – important for modernisation works – and quick installation, which translates into earlier commissioning and effective use with devices using renewable energy sources (RES), due to the use of low supply temperature and others. Its low thermal inertia allows for a temperature increase of 10°C in just 1 hour, as shown in Figure 3.13, which facilitates the control of the heating system and contributes to energy savings and improved comfort of use.

The main disadvantage was the noticeable unevenness of the temperature distribution on the tiles. It may occur with large coil spacings, but not with a spacing of 10cm and a supply temperature up to 35°C. In this case, the temperature variation has an average range of 1.6°C–2.3°C. It will only increase if we need to exert high supply temperatures or if the structure is thin when we do not use metal heat diffusers (lamellas), which help achieve uniform temperature distribution on the floor.

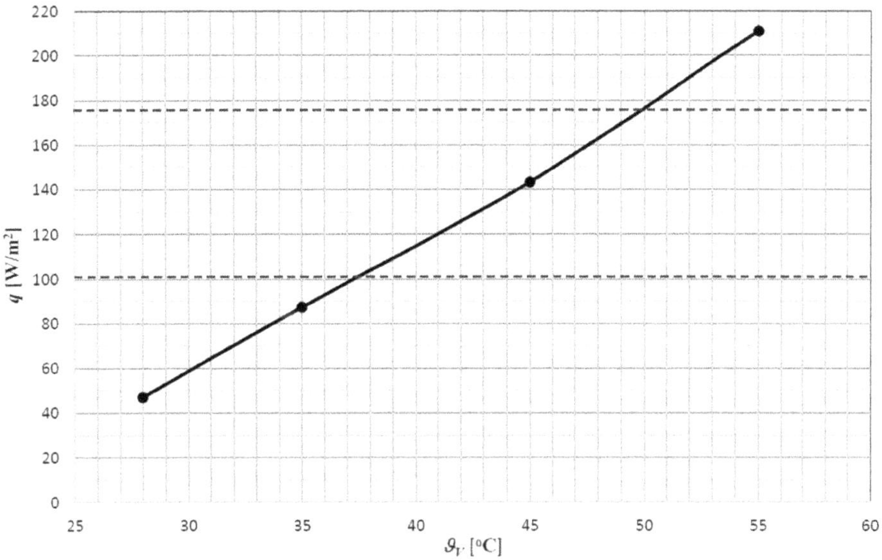

FIGURE 3.12 Dependence of the q *value* on the supply temperature ϑ_V. (Developed by M. Żukowski.)

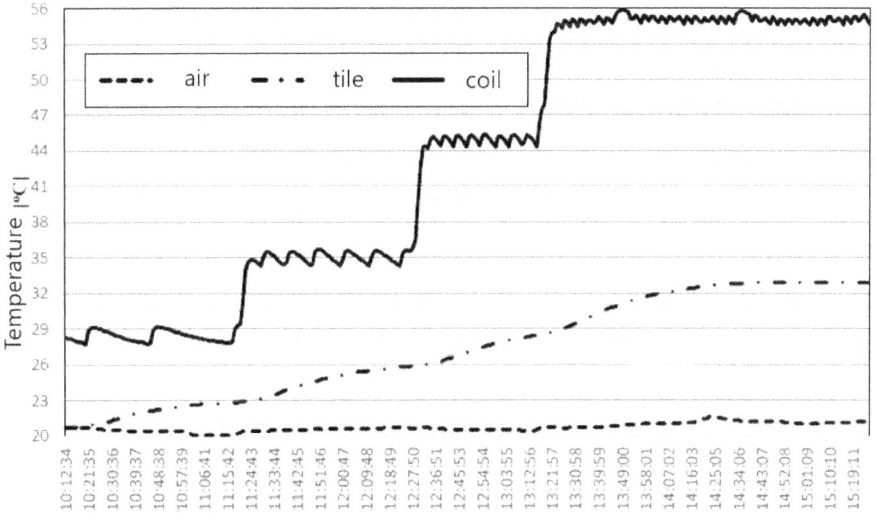

FIGURE 3.13 The heating phase of a lightweight floor radiator from [56]. (Prepared by M. Żukowski.)

In ISO 7730:2005 [34], where thermal comfort indicators are assessed, the recommended values were achieved just 3 hours after starting the ultra-thin heated floor system. In a radiant heating system, air temperature can be maintained lower than in conventional systems while achieving the same degree of thermal comfort [15].

The analysis of experimental and numerical studies [55] also shows that at different heating coil spacings (tested from 100 to 300 mm in a module every 50 mm) and at different supply temperatures, we notice a drop in air temperature when we increase the distance between the heating pipes or reduce the temperature of the power supply. This is confirmed by the authors of the article [57], on page 27, in their calculations of thermal resistance, which refer to the analysis of heat transfer in multilayer floor structures for the underfloor cooling system. The overestimation of heat transfer coefficients by 10%–30% in radiant heating/cooling, reported in standards and various studies, was noted by Cholewa et al. [58]. The reason for this state of affairs is believed to be that in studies and calculations on the heat transfer coefficient, instead of calculating the heat emitted from the surface radiator, reference was made to the heat supplied to it. If these data are overestimated and pertain to U_c and U_{total}, it may lead to higher energy consumption and thermal discomfort during operation. According to the above-mentioned authors, the calculation of U_{total}, at a height of $h = 1.1$ and 0.6 m from the heating surface, should be

$$U_{total} = 7.67\left(T_s - T_{op\ 1.1}\right)^{0.1}, \quad h = 1.1\ \text{m}$$

$$U_{total} = 8.21\left(T_s - T_{op\ 0.6}\right)^{0.1}, \quad h = 0.6\ \text{m}$$

where

T_s – floor surface temperature

T_{op} – air temperature at the optimal height h

for the cooling surface, respectively:

$$U_{total} = 5.7\left(T_s - T_{op\ 1.1}\right)^{0.1}, \quad h = 1.1\ \text{m}$$

$$U_{total} = 7.0\left(T_s - T_{op\ 0.6}\right)^{0.1}, \quad h = 0.6\ \text{m}$$

In addition, the authors of the publication recommend adopting the coefficient of thermal conductivity by radiation, in the calculations for heating, equal to 5.6 [$W/m^2{\cdot}K$] and for cooling 5.0 [$W/m^2{\cdot}K$], while for convection they propose reducing the calculations described by other authors [59] and [60] by 24%.

To investigate the main factors affecting the radiant cooling system, an experimental study [61] was conducted on the heat transfer performance and water condensation on RCPs. These models were placed horizontally in an environmental chamber with constant temperature and humidity. This model reproduced the room's natural thermal environment. The capillary panel, which supplied chilled water, was controlled by a thermostat. Three experimental models of capillary panels were made, as shown in Figure 3.14. The first model is the coil itself without any material covering. The second capillary model is covered with a metal coating, and the third is a capillary model covered with gypsum material.

The outer diameter of the capillary tube was 4.3 mm, and the inner diameter was 2.7 mm. The capillary coil consisted of ten turns, each 800 mm long. The length and width of each model were 400 mm × 400 mm. The chamber was maintained at a

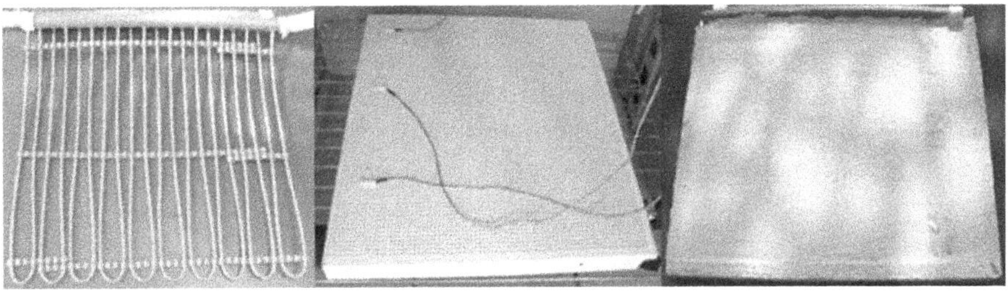

FIGURE 3.14 Three experimental models of capillary panels.

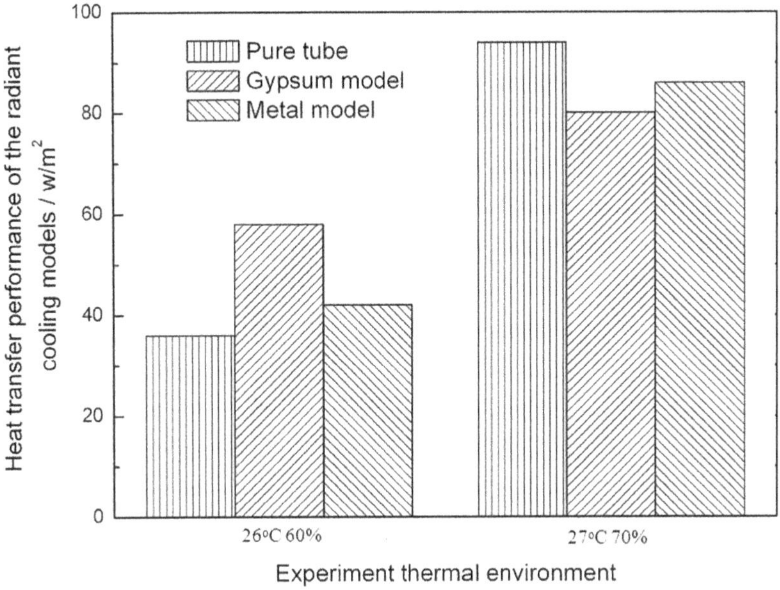

FIGURE 3.15 The heat transfer efficiency of three cooling panels under different ambient conditions.

constant temperature and relative humidity, namely 26°C and 60% humidity or 27°C and 70% humidity. The chilled water temperature on the thermostat was set to 17°C, close to the dew point inside the chamber. This way, it achieved the best heat transfer efficiency at the highest temperature difference. Figure 3.15 shows the heat transfer efficiency of three cooling panels under different ambient conditions.

At 17°C and a chilled water velocity of 0.45 m/s, the heat transfer increased from 36 to 92 W/m^2 when the ambient conditions changed from 26°C, 60% humidity to 27°C, 70% humidity. The increase in heat transfer was mainly due to the extensive water condensation on the capillary tube surface. On the other hand, there was no moisture condensation on the gypsum board, and the heat transfer increased by 36%. Among the many conclusions of this experiment, the most important is that the gypsum cooling panel had the best heat transfer efficiency compared to the metal-coated panel or the panel with only a cooling coil. It is further found that when the ambient temperature and relative humidity are 26°C and 60% RH, respectively, the supercooling degree limit (the difference between the transformation temperature and the lowest temperature indicated by the internal sensor in the initial cooling phase) for the clean panel, metal panel, and gypsum panel is 3.2°C, 5.4°C, and 6.5°C, respectively. It is worth noting that the higher the supercooling degree, the better the conditions for minimising moisture condensation on the surface. It follows that the panel covered with gypsum mortar has the best ability to avoid water condensation under identical ambient conditions. Thus, the article [61] proves that water condensation can occur under variable room conditions. For this reason, using a relatively thin hydrophobic material (gypsum, concrete, glue, etc.) on the cooling panels can ensure the safe and correct operation of the cooling system, leading to the minimisation of water condensation on the floor covering.

The same applies when describing the advantages of wall radiant heating with respect to cooling. Analysis of heat exchange in multilayer constructions flooring, presented by Li et al. [57], shows that the space between surface pipes and floors has a considerable influence on heat efficiency through radiation in systems cooling floor. It is better when these distances are smaller, using materials with the same thermal conductivity coefficient. This is also confirmed by the research of Werner-Juszczuk [62], who conducted a numerical analysis of the influence of radiator thickness on the functional properties of lightweight underfloor heating. The research work focused on optimising the thickness of aluminium sheets from 0.1 to 0.5 mm, placed in various heated floor arrangements on a 30-mm-thick EPS (expanded polystyrene) thermal insulation layer. Its parameters were

- Radiant heating with a 25-mm-thick dry screed, covered with three types of flooring – ceramic tiles, floor panels, and carpet, which are marked as A1–A3 in Figure 3.16.
- Radiant heating with a 5-mm-thick wood-like board, covered with three types of flooring – ceramic tiles, floor panels, and carpet, which are marked as C1–C3 in Figure 3.16.
- Lightweight underfloor heating without any screed, with a ceramic floor directly connected to the EPS polystyrene using a 2-mm-thick polyurethane adhesive, which is marked as B in Figure 3.16.

The heating coil spacing adopted for the analysis was 7.5, 12.5, and 15 cm, at a supply temperature of 25°C–45°C. Numerical calculations were performed using ANSYS Steady-State Thermal Solver (Workbench 19.2). A constant water velocity of 0.2 m/s was assumed, as the minimum water flow velocity for underfloor heating, required for proper operation, is 0.15 m/s. Formula (3.1) was used to calculate the heat transfer coefficient.

$$U_{total} = 8.92 \cdot (T_s - T_i)^{0.1} \tag{3.1}$$

The internal temperature on both sides of the heating surface was assumed to be 20°C. With the same test parameters, the highest floor temperatures and the highest thermal power were observed with lightweight underfloor heating. The results of numerical calculations at various heating coil spacings are presented in Figures 3.17–3.19.

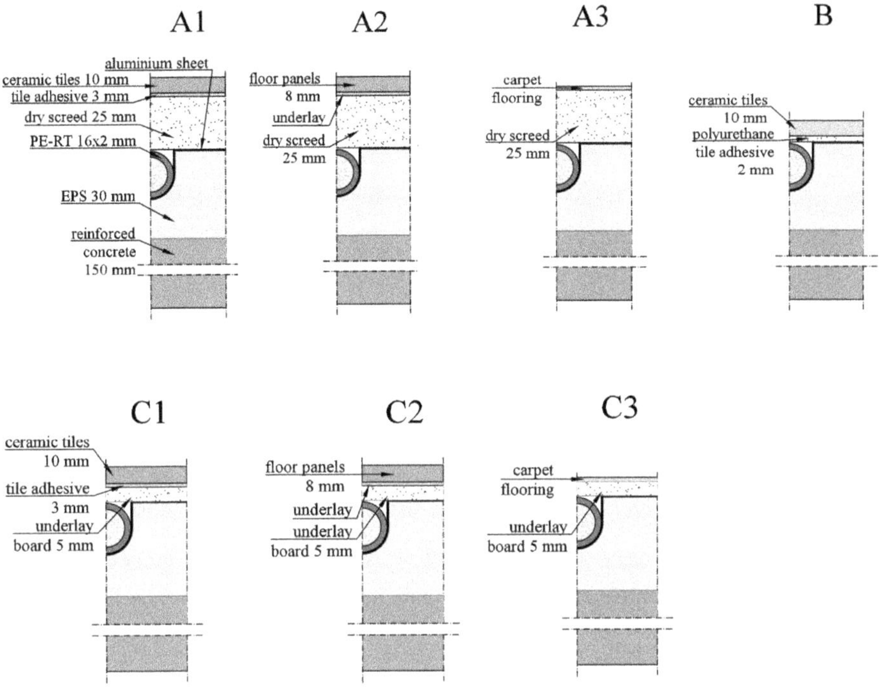

FIGURE 3.16 Models of radiant underfloor heating with and without screeds [62].

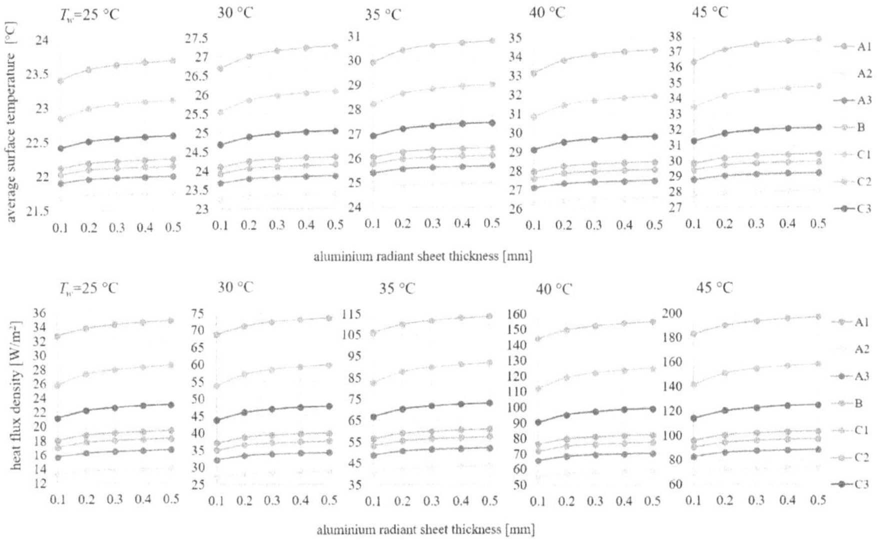

FIGURE 3.17 Dependence of different aluminium lamella thicknesses on the average floor temperature or the density of the heat flux, with different models from Figure 3.16 and different water temperatures when the heating coil spacing is 7.5 cm.

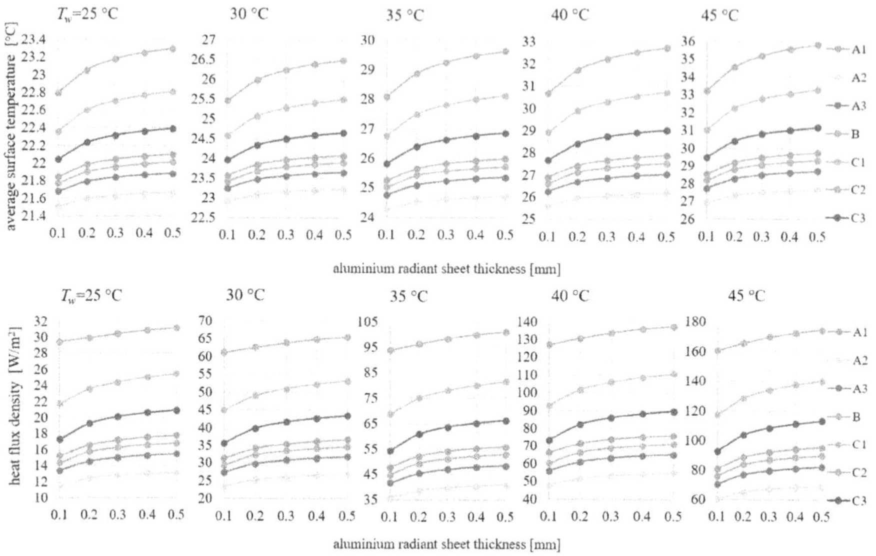

FIGURE 3.18 Dependence of different aluminium lamella thicknesses on the average floor temperature or the density of the heat flux with different radiant heating models from Figure 3.16 and different water temperatures when the heating coil spacing is 12.5 cm.

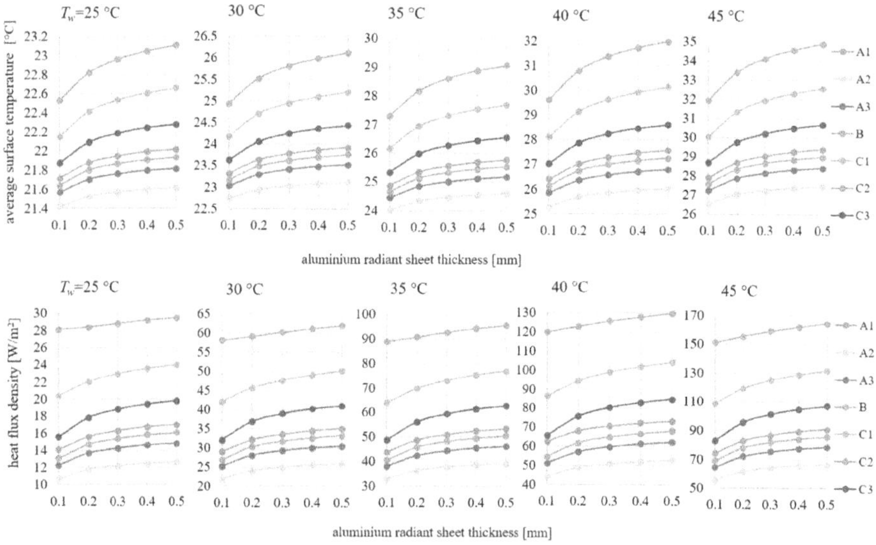

FIGURE 3.19 Dependence of different aluminium lamella thicknesses on the average floor temperature or the density of the heat flux with different radiant heating models from Figure 3.16 and different water temperatures when the heating coil spacing is 15 cm.

The basic conclusions resulting from this research are as follows:

- Optimal water temperature for all analysed solutions is 25°C–35°C when the thermal resistance of the layers above the heating coil is greater than 0.039 m^2 K/W, and 45°C water temperature is not recommended.
- The lightweight heating structure with polyurethane adhesive under ceramic tiles should be used at 25°C–30°C water temperature.
- Reducing the thickness of aluminium lamellas causes a drop in temperature on the floor surface and lower heat flux density values, and an increase in the floor temperature difference between the heating coils.
- A change in aluminium thickness in the range of 0.3–0.5 mm does not have a significant impact on the thermal power when using ceramic tiles, laminate panels, or carpet on the floor. In such circumstances, the drop in surface temperature is less than 0.5 K, and the reduction in heat power is no more than 5%.

The average temperature on the ceramic tile surface of the numerically tested lightweight underfloor heating system, described in [62] – model B, covered with polyurethane adhesive, reached the following values depending on the spacing of the heating pipes and the water supply temperature:

- at a spacing of 7.5 cm and a water temperature of 30°C – **27.2°C**,
- at a spacing of 7.5 cm and a water temperature of 35°C – **30.6°C**,
- at a spacing of 12.5 cm and a water temperature of 30°C – **26.2°C**,
- at a spacing of 12.5 cm and a water temperature of 35°C – **29.0°C**,
- at a spacing of 15 cm and a water temperature of 30°C – **25.7°C**,
- at a spacing of 15 cm and a water temperature of 35°C – **28.5°C**.

However, the average heat flux densities of this light underfloor heating system – model B, numerically tested depending on the spacing of the heating pipes and the water supply temperature, are as follows:

- at 7.5 cm spacing and 30°C water temperature – **72 W/m²**,
- at 7.5 cm spacing and 35°C water temperature – **110 W/m²**,
- at 12.5 cm spacing and 30°C water temperature – **63 W/m²**,
- at 12.5 cm spacing and 35°C water temperature – **97 W/m²**,
- at 15 cm spacing and 30°C water temperature – **60 W/m²**,
- at 15 cm spacing and 35°C water temperature – **93 W/m²**.

To verify the above data from numerical calculations of an LRHS, in which the assumed thermal resistance of the layers above the coil is 0.018 m^2 K/W, it was decided to perform our own experimental and computational tests, based on the standards and mathematical formulas available in the literature.

3.3 SUMMARY OF THE "HEAVY" AND "LIGHTWEIGHT" RADIANT HEATING AND COOLING CHARACTERISTICS

The primary goal of underfloor heating systems is to maintain a uniform room temperature and ensure thermal comfort. Thermal radiation from floors and walls reduces air temperature by approximately 2 K without compromising user comfort, while also lowering heating costs. Research has shown that the optimal floor temperature is approximately 26°C, with maximum temperatures of 29°C in the central zone and 35°C in the edge zone. In cooling systems, the minimum surface temperature should be between 18°C and 20°C. The appropriate arrangement of heating coils and the selection of materials with low thermal capacity enable a more responsive system to changes in the building's internal and external conditions. In the case of massive screeds, the systems react more slowly but retain heat longer, allowing them to take advantage of cheaper energy from the nighttime tariff. Moreover, in passive houses with a heating energy demand of less than 15 kWh/(m^2 year), LRHS ideally meet the requirements of efficiency and comfort. An essential element of radiant heating is intelligent system control, compliant with the EU ErP Directive, which mandates the use of adaptive thermostats and remote management systems. Reducing energy consumption and CO_2 emissions is a key element of the EU's energy and climate policy. Between 1990 and 2020, the EU achieved a 34% reduction in CO_2 emissions. The "Fit for 55" package being implemented aims to further reduce CO_2 emissions by 55% by 2030 and achieve climate neutrality by 2050. The development of modern heating and cooling systems that utilise renewable energy is therefore essential.

Traditional, "heavy" water systems, whose pipes are embedded in thick layers of concrete or under dry screeds, exhibit high thermal inertia. Their advantages include high thermal comfort, aesthetics (no visible radiators), quiet operation, hygienic operating conditions, adjustable temperature distribution, and compatibility with RES. Their thermal efficiency in space is approximately 100 W/m^2 for heating and 42 W/m^2 for cooling. Studies indicate that more than half of heat transfer occurs by radiation, allowing for lower indoor air temperatures. Disadvantages include a slow response to temperature changes, which makes control difficult due to high inertia.

Modern, "lightweight" heating and cooling systems without any screed can be used in all internal building partitions – on floors, walls, and ceilings, thanks to their "lightweight" construction. They are characterised by low weight, low thermal inertia, and the ability to operate at even lower supply temperatures than "heavy" systems. This makes them ideal for use with heat pumps and renewable energy installations. They enable rapid and precise control. "Lightweight" underfloor heating systems, free of heavy screeds, provide speedy heating. Studies have shown, among other things, that a 10°C temperature increase can be achieved in approximately one hour, and a screedless underfloor radiator is the most energy-efficient and works best with low-temperature heat sources. It also enables significant operating savings. Studies of a lightweight floor radiator without screeds noted that uneven temperature distribution can occur with huge coil spacing. This can be eliminated by using a dense coil arrangement, e.g., every 10 cm, or by installing metal heat diffusers. This aspect should also be the subject of further experiments and computational analyses. Numerical calculations of a "light" floor radiator indicate that at a water temperature of 30°C–35°C, a thermal output of 60–110 W/m^2 is achieved. The best results are achieved using a thin layer of polyurethane adhesive applied to the metal lamellas and a ceramic tile floor, i.e., a floor without any screed above the heating coil. In this case, the lightweight floor structure ensures rapid heating, high energy efficiency, and easy temperature control.

An alternative or complement to underfloor heating/cooling is surface wall heating or cooling. It is most often installed in modernised buildings and rooms where the floor is covered with furniture or plumbing fixtures. Advantages of lightweight wall heating include operating at low heating medium temperatures (28°C–35°C), high heat output (up to 40°C), and simple installation. Studies have shown that up to 68% of heat transfer occurs through radiation, with a total heat transfer coefficient of approximately 8–9 W/m^2·K. Compared to convectors, a wall system consumes 20%–30% less energy and emits less CO_2.

Numerous studies, calculations, and analyses indicate that lightweight surface heating and cooling systems are a future-proof solution for energy-efficient buildings. They are characterised by a short response time to temperature changes, high thermal comfort, and the ability to integrate with RES, thanks to their low supply temperature. This also makes them economical to operate. In the context of EU climate goals such as "Fit for 55", these technologies will be crucial for reducing emissions and efficiently managing energy in buildings, including heating and cooling.

ABBREVIATIONS

CO_2	carbon dioxide
CSO	Central Statistical Office
EPS	expanded polystyrene
ErP	energy-related products
EU	European Union
LRH/CS	lightweight radiant heating and cooling systems
LRHS	lightweight radiant heating system
PE-RT/AL/PE-RT	multilayer polyethylene pipe with aluminium insert
PM2.5	suspended dust with a maximum diameter of 2.5 μm
PM10	suspended dust with a maximum diameter of 10 μm
RCPs	radiant cooling panels
RES	renewable energy sources
RH	relative humidity

REFERENCES

1. Bean R., Olesen B.W., Kim K.W., History of radiant heating & cooling systems: part 1. *ASHRAE Journal*, 52 (2010), pp. 40–41.
2. EN 1264-2: 2008+A1:2012, IDT. Water-based surface-embedded heating and cooling systems. Part 2: floor heating: prove methods for the determination of the thermal output using calculation and test methods. European Committee for Standardization (CEN): Brussels, Belgium.
3. Dzierzgowski M., Rabjasz R., Wereszczyński P., *Design guidelines for electric underfloor heating*. Elektra, Warsaw, 1994. (in Polish).

4. Hao X., Zhang G., Chen Y., Zou S., Moschandreas D.J., A combined system of chilled ceiling, displacement ventilation and desiccant dehumidification. *Building and Environment*, 42 (2007), pp. 3298–3308.
5. Elektra Vademecum, Poland, edition 09/2021.
6. EN 15377-1:2008, ITD. Heating systems in buildings: design of embedded water-based surface heating and cooling systems. Part 1: determination of the design heating and cooling capacity. European Committee for Standardization (CEN): Brussels, Belgium.
7. Sankowski C., *The role of window joinery in passive construction (taking into account solar gains).* Polish Passive House Institute, Gdańsk, CS Studio 2008–2012. (in Polish).
8. Żukowski M., *Underfloor heating.* Handbook for students of Environmental Engineering. Białystok University of Technology Publishing House, Bialystok, Poland, 2009. (in Polish).
9. Januszkiewicz K., Teodorczyk T., *Control and regulation in electric room heating.* Lodz University of Technology, Department of Electrochemistry, Lodz, Poland, 1994. (in Polish).
10. Stankeviciute L., Criqui P., Energy and climate policies to 2020: the impacts of the European "20/20/20" approach. *International Journal of Energy Sector Management Emerald*, 2 (2) (2008), pp. 252–273.
11. Commission Regulation (EU) 2015/1188 of 28 April 2015 implementing Directive 2009/125/EC of the European Parliament and of the Council with regard to ecodesign requirements for local space heaters. Brussels, Belgium.
12. Koca A., Gemici Z., Topacoglu Y., Cetin G., Acet R.C., Kanbur B.B., Experimental investigation of heat transfer coefficients between hydronic radiant heated wall and room. *Energy Buildings*, 82 (2014), pp. 211–221.
13. Strzeszewski M., Electric heating: reducing low emissions and counteracting smog. *Hot News*, 1(28), (2017), pp. 1–2. (in Polish).
14. Regulation of the Minister of Transport, Construction and Maritime Economy of July 5, 2013 amending the regulation on the technical conditions to be met by buildings and their location. (in Polish).
15. Rhee K.N., Kim K.W., A 50 year review of basic and applied research in radiant heating and cooling systems for the built environment. *Building and Environment*, 91 (2015), pp. 166–190.
16. Bojić M., Cvetković D., Marjanović V., Blagojević M., Djordjević Z., Performances of low temperature radiant heating systems. *Energy Buildings*, 61 (2013), pp. 233–238.
17. Babiak J., Olesen B.W., Petras D., *Low temperature heating and high temperature cooling.* REHVA guidebook no. 7, Rehva, Ixelles (2009).
18. Olesen B.W., Radiant floor heating in theory and practice. *ASHRAE Journal*, 44 (2002), pp. 19–26.
19. Werner-Juszczuk A.J., Analysis of the use of radiant floor heating as a cooling system. *Proceedings*, 16 (23) (2019). https://doi.org/10.3390/proceedings2019016023.
20. ISO 11855-1:2012(E). *Building environment design: design, dimensioning, installation and control of embedded radiant heating and cooling systems. Part 1: definition, symbols, and comfort criteria.* International Organization for Standard, Genève, Switzerland, 2012.
21. ANSI/ASHRAE. *Standard 55 Thermal environmental conditions for human occupancy.* American Society of Heating, Refrigerating, and Air-Conditioning Engineers, Atlanta, GA, 2010.
22. Feustel H.E., Stetiu C., Hydronic radiant cooling: preliminary assessment. *Energy Buildings*, 22 (1995), pp. 193–205.
23. Stetiu C., Energy and peak power savings potential of radiant cooling systems in US commercial buildings. *Energy Buildings*, 30 (1999), pp. 127–138.

24. Peng P., Gong G., Mei X., Liu J., Wu F., Investigation on thermal comfort of air carrying energy radiant air-conditioning system in south-central China. *Energy Buildings*, 182 (2019), pp. 51–60. https://doi.org/10.1016/j.enbuild.2018.10.020
25. Zhou X., Liu Y., Luo M., Zhang L., Zhang Q., Zhang X., Thermal comfort under radiant asymmetries of floor cooling system in 2h and 8h exposure durations. *Energy Buildings*, 188–189 (2019) 98–110. https://doi.org/10.1016/j.enbuild.2019.02.009
26. De Carli M., Tonon M., Effect of modelling solar radiation on the cooling performance of radiant floors. *Solar Energy*, 85 (2011), pp. 689–712. https://doi.org/10.1016/j.solener.2010.12.012
27. Yin Y., Wang R.Z., Zhai X.Q., Ishugah T.F., Experimental investigation on the heat transfer performance and water condensation phenomenon of radiant cooling panels. *Building and Environment*, 71 (2014) pp. 15–23. https://doi.org/10.1016/j.buildenv.2013.09.016
28. Sastry G., Rumsey P., VAV vs. radiant: side-by-side comparison. *ASHRAE Journal*, 56 (2014), pp. 16–24.
29. Tantiwichien A.-U.-W., An experimental and simulated study on thermal comfort. *International Journal of Engineering and Technology*, 5 (2013), pp. 177–180. https://doi.org/10.7763/ijet.2013.v5.536
30. Yau Y.H., Hasbi S., Field analysis of indoor air quality in high rise and low rise green offices with radiant slab cooling systems in Malaysia. *Indoor and Built Environment*, 24 (2015), pp. 174–184. https://doi.org/10.1177/1420326X13506130
31. ASHRAE, *ASHRAE handbook: HVAC systems and equipment.* American Society of Heating, Refrigerating, and Air Conditioning Engineers, Atlanta, GA, 2016.
32. Zhang F., Guo H.-A., Liu Z., Zhang G., A critical review of the research about radiant cooling systems in China. *Energy Buildings*, 235 (2021), p. 110756.
33. Tye-Gingras M., Gosselin L., Comfort and energy consumption of hydronic heating radiant ceilings and walls based on CFD analysis. *Building and Environment,* 54 (2012), pp. 1–13.
34. ISO 7730. *Ergonomics of the thermal environment-analytical determination and interpretation of thermal comfort using calculation of the PMV and PPD indices and local thermal comfort criteria.* International Organization for Standardization, Geneva, Switzerland, 2005.
35. Song G.S., Could sperm quality be affected by a building environment? A literature review. *Building and Environment*, 45 (2010), pp. 936–943.
36. Cvetković D., Bojić M., Optimization of thermal insulation of a house heated by using radiant panels. *Energy Buildings*, 85 (2014), pp. 329–336.
37. Le Dréau J., Heiselberg P., Sensitivity analysis of the thermal performance of radiant and convective terminals for cooling buildings. *Energy Buildings*, 82 (2014), pp. 482–491.
38. Kim K.W., Olesen B., Radiant heating and cooling systems: part 1. *ASHRAE Journal*, 57 (2015), pp. 28–37.
39. Karpiesiuk J., Chyży T., Comparative analysis of the thermal efficiency of wall surface radiators with a lightweight dry structure. Post-conference monograph as part of the *1st International Conference Entitled Current Research Problems of Materials, Technologies and Organization of Construction in a Cross-Border Perspective*, Białystok University of Technology, June 2016, Białystok. (in Polish).
40. Technical guide part 1 Upnor. Surface heating/cooling, Poland, 2012. (in Polish).
41. EN 12831-1:2017, ITD. Energy performance of buildings: method for calculation of the design heat load. Part 1: space heating load, Module M3-3. European Committee for Standardization (CEN): Brussels, Belgium.
42. EN 15377-2:2008. Heating systems in buildings: design of embedded water based surface heating and cooling systems. Part 2: design, dimensioning and installation. European Committee for Standardization (CEN): Brussels, Belgium.

43. EN 15377-3:2007. Heating systems in buildings: design of embedded water based surface heating and cooling systems. Part 3: optimizing for use of renewable energy sources. European Committee for Standardization (CEN): Brussels, Belgium.
44. EN 1264-4:2009. Water based surface embedded heating and cooling systems. Part 4: installation. European Committee for Standardization (CEN): Brussels, Belgium.
45. Surface heating and cooling guide. KAN-therm system. PL, Poland, 11/2015. (in Polish).
46. Karpiesiuk J., *Ultra-thin radiator. Heat from the floor without screeds (2)*, Magazyn Instalatora, 02/2016, pp. 20–21. Poland. (in Polish).
47. EN 1264-5:2021, ITD. Water-based surface embedded heating and cooling systems. Part 5: determination of the thermal output for wall and ceiling heating and for floor, wall, and ceiling cooling. European Committee for Standardization (CEN): Brussels, Belgium.
48. Causone F., Corgnati S.P., Filippi M., Olesen W.B., Experimental evaluation of heat transfer coefficients between radiant ceiling and room. *Energy Buildings*, 41 (2009), pp. 622–628.
49. Todorovic B., *Design of installations for central heating.* Masinski Faculty Beograd, Beograd, 2004.
50. Acikogoz O., A novel evaluation regarding the influence of surface emissivity on radiative and total heat transfer coefficients in radiant heating systems by means of the theoretical and numerical methods. *Energy Buildings*, 102 (2015), pp. 105–116.
51. Bojić M., Cvetković D., Miletić M., Malešević J., Boyer H., Energy, cost and CO_2 emissions between radiant wall panel systems and radiator systems. *Energy Buildings*, 54 (2012), pp. 496–502.
52. Crawley D.B., Lawrieb L.K., Winkelmannc F.C., Buhlc W.F., Huangc Y.J., Pedersend C.O., Strandd R.K., Liesend R.J., Fishere D.E., Wittef M.J., Glazerf J., EnergyPlus: creating a new generation building energy simulation program. *Energy Buildings*, 33 (2001), pp. 443–457.
53. EN 1264-5:2008, ITD. Water-based surface embedded heating and cooling systems. Part 5: determination of the thermal output for wall and ceiling heating and for floor, wall, and ceiling cooling. European Committee for Standardization (CEN): Brussels, Belgium.
54. Karpiesiuk P., *Research on the temperature field during surface heating.* Engineering Diploma Thesis, Faculty of Construction and Environmental Engineering, Department District Heating, Heating and Ventilation, 2015. (in Polish).
55. Dongliang Z., Ning C., Zijie W., Experimental and numerical analysis of lightweight radiant floor heating system. *Energy Buildings*, 61 (2013), pp. 260–266.
56. Żukowski M., Karpiesiuk P., Test results of a radiant radiator with a very low height. *Instal*, 10 (2015), pp. 38–41. (in Polish).
57. Li Q.-Q., Chen C., Zhang Y., Lin J., Ling H.-S., Simplified thermal calculation method for structure in radiant floor cooling system. *Energy Buildings*, 74 (2014), pp. 182–190.
58. Cholewa T., Rosinski M., Spik Z., Dudzinska M.R., Siuta-Olcha A., On the heat transfer coefficients between heated/cooled radiant floor and room. *Energy Buildings*, 66 (2013), pp. 599–606.
59. Awbi H.B., Hatton A., Natural convention from heated room surfaces. *Energy Buildings*, 30 (1999), pp. 233–244.
60. Min T.C., Schutrum L., Parmelee G., Vouris J., Natural convection and radiation in a panel heated room. *ASHRAE Transactions,* 62 (1956), pp. 153–160.
61. Yin Y.L., Wang R.Z., Zhai X.Q., Ishugah T.F., Experimental investigation on the heat transfer performance and water condensation phenomenon of radiant cooling panels. *Building and Environment*, 71 (2014), pp. 15–23.
62. Werner-Juszczuk A.J., The influence of the thickness of an aluminum radiant sheet on the performance of the lightweight floor heating. *Journal of Building Engineering*, 44 (2021), p. 102896.

4 Laboratory Research, Thermal Power Calculations, and User Comfort of a Lightweight Radiant Heating System (LRHS)

The experimental studies of the lightweight radiant heating system (LRHS) without screeds were based on the thermal power and maximum water supply temperature formulas from the NT VVS 127:2001 standard. This standard is the only one worldwide that applies exclusively to lightweight underfloor heating systems (no screed requirement) using metal heat-dissipating elements. Other lightweight underfloor heating solutions, lacking various types of conductive panels or lamellas, should be designed solely through experimental research. Therefore, research models were prepared, and thermal efficiency calculations were performed for both a heated floor system with metal heat-dissipating lamellas and one without them, using only ceramic tile floors. Additionally, the maximum supply temperature for models with lamellas was calculated, and user comfort was assessed using practical experience criteria. Due to the lack of thermal efficiency formulas for lightweight underfloor heating without lamellas, proprietary formulas were developed based on formulas from the NT VVS 127:2001 standard. The entire LRHS calculation with ceramic tile flooring is summarised in the tables. The many advantages of screedless LRHS, as described in the summary of this chapter, the main ones being greater thermal efficiency than traditional underfloor heating systems and more efficient temperature control, may make this construction widely used in the coming years. Radiant heating installed on the floor is typically achieved using coils covered with a wet cement screed. Lightweight surface heating systems are not yet as popular. Numerous studies show that radiant heat transfer, which is over 50%, improves comfort and reduces heating costs when using lower supply temperatures. The optimal floor temperature is around 26°C for heating and no less than 18°C for cooling. The low surface thermal inertia, due to the absence of screeds, allows better control of the heating system and higher heat power. Surface heating systems are most effective on the floor, while cooling systems are most effective on the ceiling. Both systems are more effective when combined with low-temperature heat sources, without any screed. Metal lamellas provide an effective means of distributing heat or cooling. The lack of

DOI: 10.1201/9781003679004-4

lamellae can be compensated for by denser coil spacing. Numerous studies, calculations, and analyses indicate that lightweight surface heating and cooling systems are a future-proof solution for energy-efficient buildings due to their short response time to temperature changes, higher thermal efficiency than "heavy" systems, and more effective integration with renewable energy sources thanks to their lower supply temperature compared to traditional surface systems.

In 2017, experimental tests were performed in accordance with the NT VVS 127:2001 standard [1] for lightweight radiant underfloor heating systems using metal heat-dissipating elements. Non-standard experiments were also performed without their use. According to the provisions of the standard [1], it can also be extended when designing electric underfloor heating. Other solutions for radiant heating, without using conductive boards or lamellas, can only be designed through experimental testing. When preparing the test stand, various heating coil spacings were used. One of the experiment's goals was to address the defect in lightweight radiant heating system (LRHS) observed during the engineering work described in [2] during the examination of the thermal parameters of a lightweight radiator without heat-dissipating lamellas. It was questioned whether this type of heating system, without screeds, would achieve a comfortable temperature distribution on the floor surface by increasing the spacing between the heating coils when there are no metal dissipative layers. The next issue was to determine the thermal efficiency if we used a heat-conducting aluminium foil glued directly to the insulating board.

The second reason for experimenting on a lightweight radiant floor was to verify the numerical calculations of thermal power made by Werner-Juszczuk [3] using the ANSYS computer program. The performed numerical calculations analysed, among other factors, the influence of the aluminium heat spreader's thickness on the operational properties of a lightweight underfloor radiator. The modelled lightweight radiator surface was powered by water from a 16-mm-diameter heating coil, placed on 30-mm-thick expanded polystyrene (EPS) thermal insulation with built-in aluminium lamellas 0.1–0.5 mm thick. This structure was covered with a 2-mm-thick polyurethane adhesive and a 10-mm-thick ceramic tile. A detailed description of the guidelines, charts, and conclusions of numerical calculations can be found in the publication [3] and Chapter 3.

It was decided to calculate and compare the thermal power of lightweight floor radiators using both theoretical and experimental methods. Theoretical calculations of thermal power were performed using existing standards and the author's original calculation formulas (applied only to LRHS without lamellas). The theoretical thermal power calculations from the formulas are based on the heating coil supply temperature. Experimental calculations are based on formulas that account for the average flooring temperature when calculating the thermal power.

4.1 DESCRIPTION OF MEASUREMENT STANDS AND MEASUREMENT METHODOLOGY

To verify the results and conclusions of the engineering work [2] and those resulting from numerical calculations [3], regarding only the lightweight underfloor radiator, three test stands were created. Each of them was divided into two equal parts, and the heating pipes were arranged at different spacings of 12.5–20 cm, as described in the article by Karpiesiuk and Chyży [4]. The heat source was electric heating cables

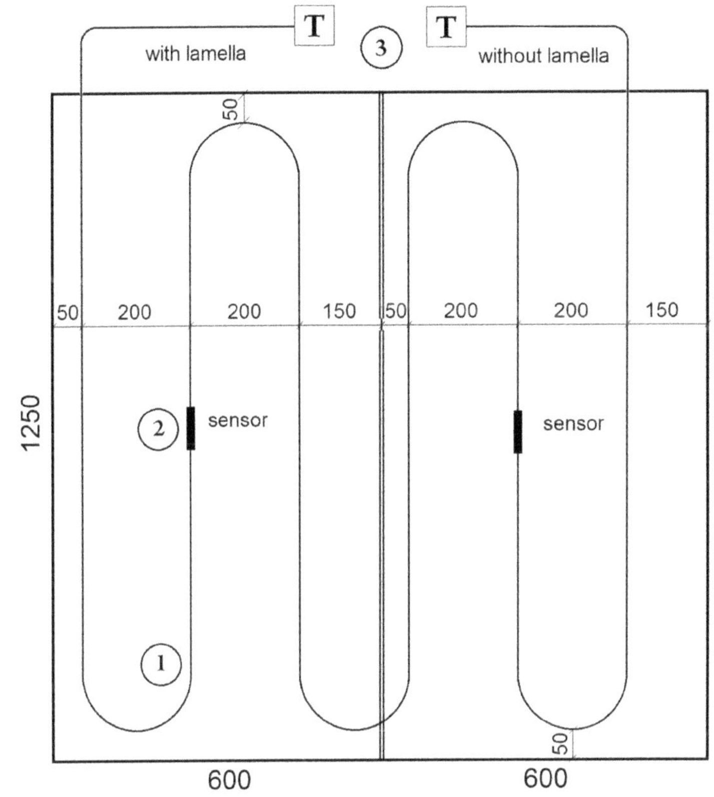

FIGURE 4.1 Diagrams of measurement stands. 1. Heating coil, 2. Temperature sensor, 3. Thermostat for measuring the air temperature and power supply to the coil.

installed within the heating coil, in accordance with the test conditions specified in the standard [1]. The heating cables have a stable temperature throughout the entire coil, so it can be assumed that there are no differences between the inlet and outlet temperatures of the heating pipes. Therefore, it was sufficient to place thermistors (sensors) for temperature measurement, located according to the diagram shown in Figure 4.1. On one side of these three stands, a thermal insulation system with aluminium lamellas was installed as a heat-conducting coating to dissipate heat. This part was covered with waterproof polyurethane adhesive. On the other side, no heat spreaders were used, and the heating pipes were directly inserted into grooves carved in the thermal insulation and covered with cement adhesive mortar. The whole structure was covered with ceramic tiles (Figures 4.2 and 4.3).

Diagrams of the measurement stands and their vertical sections are presented in Figures 4.4 and 4.1. The aluminium layer inside the pipe further ensured uniform heat dissipation from the heating cable. To measure the temperature on the heating pipe and in the room, electronic programmable thermostats ST-500 were used, with a measurement accuracy of ±0.5°C and two sensors – one for air temperature and one for floor temperature. The average air temperature of the laboratory hall near the research models and the surface temperature of the partitions were 20°C.

FIGURE 4.2 Three test stands, covered with ceramic tiles.

FIGURE 4.3 Test stands with and without aluminium lamellas, covered with cement adhesive (on the left) or polyurethane adhesive (on the right).

Temperature measurements directly on the heating surface were scanned using a FLIR I40 thermal imaging camera and a TN1 pyrometer from OJ Electronics, with the thermal sensitivity and temperature measurement accuracy the same as for the thermal imaging camera. In the experiment, the basic conditions for preparing test stands were met in accordance with the standard NT VVS 127:2001 [1], point 7. It was decided to perform calculations and compare the results for heat flux density (thermal power) using standard theoretical formulas and the original author's formulas.

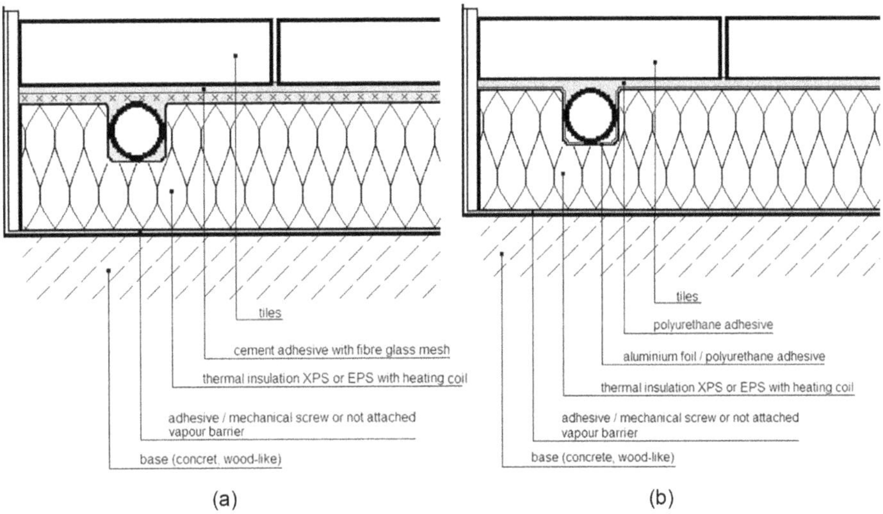

FIGURE 4.4 Vertical cross-section of the measurement stands, covered with (a) cement adhesive and (b) polyurethane adhesive.

4.2 METHODOLOGY FOR CALCULATING THE THERMAL POWER AND MAXIMUM TEMPERATURE IN THE HEATING PIPE OF A LIGHTWEIGHT RADIANT HEATING SYSTEM WITH LAMELLAS

Due to the lack of heat flux meters in experimental studies, the heat flux density emitted from the tested surface was determined using formula No. 7, point 5.3, as recommended in the NT VVS127:2001 standard [1]. The formula has been corrected by the author of this book due to an error. The author, analysing the formula of the Norwegian standard for the balanced heat transfer coefficient K_H, noticed that, in the denominator, instead of the maximum temperature of the medium in the heating coil θ_H^{max}, the maximum allowable floor surface temperature θ_F^{max} should be used.

Hence, the appropriate form of the formulas from the above-mentioned standard [1] for calculating the theoretical heat flux density should have the form (4.1) and (4.2). This is proven by the calculations' results and their comparison with other theoretical formulas for calculating the heat flux density on the floor surface, as well as by the results of the experimental tests. In addition, the incorrect entry in the KH formula is confirmed by the calculations of the efficiency of lightweight underfloor heating according to the NORDTEST NT VVS 127 method carried out by Werner-Juszczuk [5]. The article [5] examined the suitability of the Nordest Method NT VVS 127 for determining the operational parameters of lightweight underfloor heating, with a dry screed 20 or 25 mm thick on which ceramic tiles were glued. The insulating base for the heating coil was 30-mm-thick EPS covered with 0.3 mm aluminium lamella. The entire lightweight floor was placed on a 150-mm-thick reinforced concrete ceiling. For such a model, the methodology for determining surface temperature and heat flux density using the Nordest NT VVS 127 method is presented. Calculations were made for lightweight underfloor heating with a dry screed and variable water temperature. The calculation results obtained using the Nordest method were compared with numerical calculations using the ANSYS Steady-State Thermal Solver and in accordance with the PN-EN 1264 standard. There were no proposed corrections in the formulas by the author of this paper. The article's summary states that there are significant differences between the results obtained from numerical calculations and those from the Nordest Method NT VVS 127. The temperature error is up to 5.68 K, and the heat flux density error is up to 55.5%. Smaller differences in the results for these parameters were observed when comparing the PN-EN 1264 standard and numerical calculations. Here, the temperature error did not exceed 0.2 K, and the maximum heat flux density error was 8.5%. Therefore, it was found that the PN-EN 1264 standard is more suitable for determining the operational parameters of lightweight underfloor heating with dry screed than the Nordest NT VVS 127 method.

To determine the design thermal power q_i, the basic formulas marked as (4.1), (4.2), (4.3), (4.3a), (4.4), and the auxiliary formulas presented below (without numerical designations) should be used to determine these four main formulas:

$$q_i = K_{Hl} \cdot D\theta_H \ [W/m^2] \tag{4.1}$$

where

q_i – computational heat flux density (thermal efficiency) on the floor surface $D\theta_H = \theta_H - \theta_i$

θ_H– heating coil supply temperature [°C]

θ_i – room temperature [°C]

$$K_{Hl} = \frac{q_i^{max}}{\theta_F^{max} - \theta_i} \ [W/K\,m^2] \tag{4.2}$$

where

K_{Hl} – equivalent heat transmission coefficient with lamellas

θ_F^{max} – maximum permissible floor surface temperature (in the middle zone equal to 29°C)

q_i^{max} – maximum heat flux density depending on the heating coil spacing [W/m^2]

θ_i – room temperature [°C]

$$q_i^{max} = 8.92\left(\bar{\theta}_F^{max} - \theta_i\right)^{1.1} \ [W/m^2] \tag{4.3}$$

where

$\bar{\theta}_F^{max} - \theta_i$ is calculated from the formula:

$$\bar{\theta}_F^{max} - \theta_i = k_{CL}\left(\theta_F^{max} - \theta_i\right) [°C] \tag{4.3a}$$

$\bar{\theta}_F^{max}$ – maximum permissible average floor temperature [°C]

θ_F^{max} – maximum permissible floor surface temperature [°C]

k_{CL} – thermal conductivity coefficient of the dissipative layer (lamellas)

θ_i – room temperature [°C]

k_{CL} coefficient is according to the following formulas:

$$k_{CL} = \frac{L_U + 2L_{fin} \cdot k_{fin} + 0.01L \cdot k_{fin}^{min}}{T} \tag{4.4}$$

where

L_U – external diameter of the heating pipe [m]

L_G – gap between conductive layers [m]

T – spacing of heating pipes (coils) [m]

$k_{fin}^{min} = \dfrac{1}{\cosh\left(\dfrac{L_{fin}}{\ell}\right)}$ – the minimum coefficient of thermal conductivity of devices (lamella), taking into account the gaps between them

$$L_{fin} = \frac{T - L_U - L_G}{2}$$

L_{fin} – lamella width (horizontal part of the heat-conducting device as a metal lamella) [m]

$$k_{fin} = \frac{\ell}{L_{fin}} \tanh\left(\frac{L_{fin}}{\ell}\right)$$

k_{fin} – thermal conductivity coefficient of devices (lamella)

$$\ell = \sqrt{\frac{S_{WL} \cdot \lambda_{WL}}{\frac{1}{R_i} + \frac{1}{R_e}}}$$

ℓ – characteristic length of the conductive layer (lamella) [m]

S_{WL} – thickness of the conductive layer (lamella) [m]

λ_{WL} – thermal conductivity of the conductive layer [W/m·K]

R_i – thermal resistance of the layer conducting heat to the heated room [m^2 K/W]

R_e – thermal resistance under the layer that conducts heat to the room below or on **the** outside [m^2 K/W]

$$R_e = R_{e,0} - 2 \cdot \Delta R_e$$

$$\Delta R_e = \frac{d}{\lambda_{ins}} \cdot \frac{b}{T}$$

d – groove depth [m]

b – groove width [m]

λ_{ins} – thermal conductivity coefficient of thermal insulation [W/m·K]

T – spacing of heating pipes (coils) [m]

$$R_{e,0} = R_u + R_{se} = \frac{d_k}{\lambda_k} + R_{se}$$

R_{se} – heat transfer resistance under the floor surface [m^2 K/W]

R_u – thermal insulation resistance [m^2 K/W]

d_k – thermal insulation thickness [m]

λ_k – thermal conductivity coefficient of thermal insulation [W/m·K]

$$R_i = R_o + R_{con,i} + R_{si}$$

$R_{con,i}$ – resistance for conductive layers, carefully shaped and permanently glued with a layer of insulation [m² K/W]

$R_o = \frac{d_i}{\lambda_i}$ – thermal resistance of floor layers above the conductive layer [m² K/W]

d_i – thickness of individual floor layers above the conductive layer [m]

λ_i – thermal conductivity coefficient of individual floor layers above the conductive layer [W/(m·K)]

$R_{si} = \frac{1}{\alpha_i}$ – heat transfer resistance on the floor surface [m² K/W]

α_i – heat transfer coefficient on the floor surface [W/m²·K]

To calculate the maximum permissible temperature of the medium in the heating pipe θ_H^{max}, we use formulas (4.5) and (4.6) as basic ones and the auxiliary formulas presented below (without numerical markings):

$$\theta_H^{max} = \theta_i + q_i^{max}\left(R_i + \frac{1}{\eta}\text{: times } R_{HC}\right)[°C] \tag{4.5}$$

$$R_{HC} = T \cdot R'_R + T \cdot R'_{R,con} \frac{T}{2} \cdot R'_U + R_{CL}\ [m^2 \cdot °C/W] \tag{4.6}$$

where

θ_H^{max} – maximum temperature of the medium in the heating coil [°C]

R_{HC} – total thermal resistance between the heat source and the conductive layer [m²·°C/W]

q_i^{max} – maximum heat flux density [W/m²]

θ_i – room temperature [°C]

R_i – thermal resistance of the layer conducting heat to the heated room [m²·°C/W]

R'_R – total thermal resistance of the coil wall [m²·°C/W]

$R'_{R,con}$ – heat transfer resistance of the coil wall [m²·°C/W]

R'_U – linear resistance [m²·°C/W]

R_{CL} – fictitious thermal resistance of the conductive layer [m²·°C/W]

T – spacing of heating pipes (coils) [m]

$$\eta = \frac{R_e}{R_i + R_e}$$

η – efficiency of the underfloor heating system

R_e – thermal resistance under the layer that conducts heat to the room below or outside [m² K/W]

$$R'_U = \frac{0.008}{d_a}$$

$$R'_{R,con} = \frac{0.01}{d_a}$$

$$R'_R = \frac{1}{2\pi\lambda_R} \ln\left(\frac{d_o}{d_{ip}}\right)$$

d_a – pipe diameter [m]
d_o – external pipe diameter [m]
d_{ip} – internal pipe diameter [m]
λ_R – thermal conductivity of the heating pipe wall [W/m°C]

$$R_{CL} = \frac{1}{\frac{1}{R_i} + \frac{1}{R_e}} \left(\frac{1}{k_{CL}} - 1\right)$$

k_{CL} – thermal conductivity coefficient of the dissipative layer (lamellas), calculated from formula (4.4)

Heat exchange in the structure of a lightweight heated floor without screeds depends on many component factors. The complex interaction of convection, radiation, and heat conduction between the pipe and the surrounding materials makes it difficult to describe and calculate the thermal efficiency of this phenomenon, according to [6]. In simple terms, these relationships can be reduced to three thermal resistances shown in Figure 4.5.

4.2.1 Theoretical Method for Calculating the Heat Flux Density, Using the NORDTEST NT VVS 127 Method with Lamellas

The calculation example of thermal power concerns experimental tests based on the adopted materials and layers of lightweight flooring, as described in point (4.2), based on the formulas written in point (4.2). To calculate the theoretical density of the heat flux q_i emitted from the tested surface, the following calculations were made at a specific spacing of heating pipes.

Theoretical calculations of the maximum heat flux density with the lamella, depending on the heating coil distance, according to formula (4.3), based on the NT VVS127:2001 standard [1]:

$$q_i^{max} = 8.92\left(\overline{\theta}_F^{max} - \theta_i\right)^{1.1} \text{ [W/ m}^2\text{]}$$

$$\overline{\theta}_F^{max} - \theta_i = k_{CL}\left(\theta_F^{max} - \theta_i\right) [°C]$$

$$k_{CL(12.5)} = \frac{L_U + 2L_{fin} \cdot k_{fin} + 0.01L_G \cdot k_{fin}^{min}}{T} = \frac{0.016 + 2 \cdot 0.0545 \cdot 0.766}{0.125} = 0.796$$

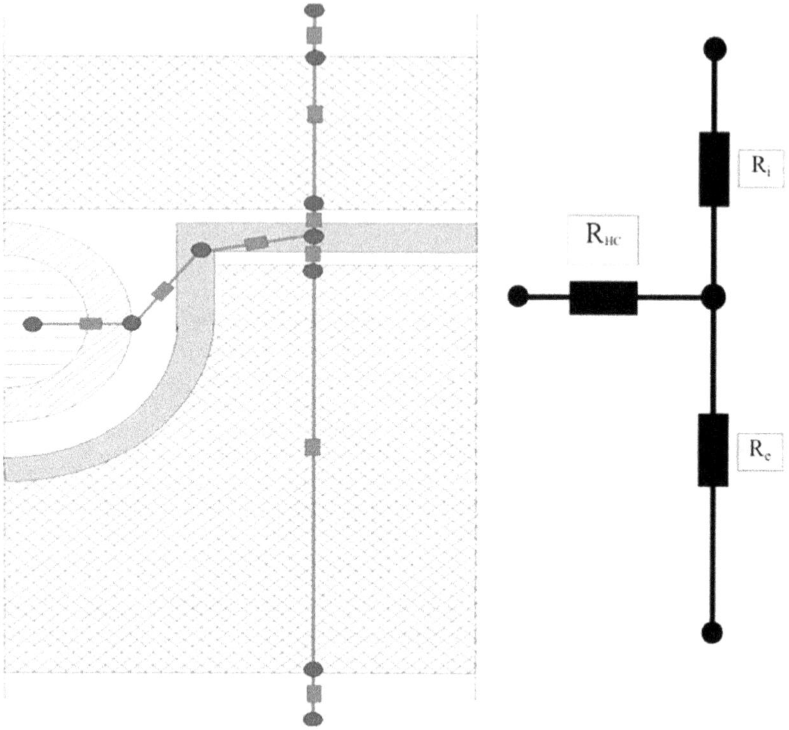

FIGURE 4.5 Heat exchange in the floor structure with heat-conducting lamellas, where R_{HC} – total thermal resistance between the heat medium and the conductive layer, R_i – thermal resistance of the layer conducting heat to the heated room, and R_e – thermal resistance under the layer that conducts heat to the room below or outside [$m^2 K/W$].

$$k_{CL(15)} = \frac{L_U + 2L_{fin} \cdot k_{fin} + 0.01 L_G \cdot k_{fin}^{min}}{T} = \frac{0.016 + 2 \cdot 0.067 \cdot 0.703}{0.15} = 0.735$$

$$k_{CL(20)} = \frac{L_U + 2L_{fin} \cdot k_{fin} + 0.01 L_G \cdot k_{fin}^{min}}{T} = \frac{0.016 + 2 \cdot 0.092 \cdot 0.573}{0.20} = 0.607$$

Substituting the k_{CL} values at specific coil spacing into the formula, we have

In the middle zone:

$$\overline{\theta}_F^{max} = \theta_i = k_{CL}\left(\theta_F^{max} - \theta_i\right)$$

$$0.796(29-20) = 7.164\ °C \text{ (for a pipe spacing of } \mathbf{12.5\ cm}\text{)}$$

$$0.735(29-20) = 6.615\ °C \text{ (for a pipe spacing of } \mathbf{15\ cm}\text{)}$$

$$0.607(29-20) = 5.463\ °C \text{ (for a pipe spacing of } \mathbf{20\ cm}\text{)}$$

In the edge zone:

$$\bar{\theta}_F^{max} - \theta_i = k_{CL}\left(\theta_F^{max} - \theta_i\right)$$

$$0.796(35-20) = 11.94°C \text{ (for a pipe spacing of } \mathbf{12.5\ cm}\text{)}$$
$$0.735(35-20) = 11.03°C \text{ (for a pipe spacing of } \mathbf{15\ cm}\text{)}$$
$$0.607(35-20) = 9.11°C \text{ (for a pipe spacing of } \mathbf{20\ cm}\text{)}$$

The following are the calculations of the individual components of the formula for the thermal conductivity coefficient of the dissipative layer (lamellas) k_{CL}, with appropriate heating coil spacing, leading to its determination:

$$L_U = 0.016\ m$$

$$L_G = 0.00\ m$$

$$L_{fin} = \frac{T - L_U - L_G}{2} = \frac{0.125 - 0.016 - 0}{2} = 0.0545 \text{ (for a pipe spacing of 12.5 cm)}$$
$$= 0.067 \text{ (for a pipe spacing of 15 cm)}$$
$$= 0.092 \text{ (for a pipe spacing of 20 cm)}$$

$$k_{fin} = \frac{\ell}{L_{fin}} \tanh\left(\frac{L_{fin}}{\ell}\right) = \frac{0.0553}{0.0545} \tanh 0.9855 = 1.015 \cdot 0.755$$
$$= 0.766 \text{ (for a pipe spacing of } \mathbf{12.5\ cm}\text{)}$$
$$= \frac{0.057}{0.067} \tanh 1.175 = 0.851 \cdot 0.826 = 0.703 \text{ (for a pipe spacing of } \mathbf{15\ cm}\text{)}$$
$$= \frac{0.057}{0.092} \tanh 1.614 = 0.620 \cdot 0.924 = 0.573 \text{ (for a pipe spacing of } \mathbf{20\ cm}\text{)}$$

Calculations of hyperbolic tangent according to http://calcoolator.pl/tangens_kalkulator.html

$$\ell_{xps} = \sqrt{\frac{S_{WL} \cdot \lambda_{WL}}{\frac{1}{R_i} + \frac{1}{R_e}}} = \sqrt{\frac{0.00009 \cdot 200}{\frac{1}{0.21} + \frac{1}{1.26}}} = \sqrt{\frac{0.018}{4.76 + 0.79}}$$
$$= 0.057 \text{ (for pipe spacings of 15 and 20 cm)}$$

$$\ell_{xps} = \sqrt{\frac{S_{WL} \cdot \lambda_{WL}}{\frac{1}{R_i} + \frac{1}{R_e}}} = \sqrt{\frac{0.00009 \cdot 200}{\frac{1}{0.21} + \frac{1}{1.26}}} = \sqrt{\frac{0.018}{4.76 + 1.13}}$$

$$= 0.0553 \left(\text{for a pipe spacing of } 12.5 \text{ cm}\right)$$

where

$S_{WL} = 0.00009$ [m] – thickness of the conductive layer (aluminium lamella)
$\lambda_{WL} = 200$ [W/(m·K)] – thermal conductivity of the conductive layer (**aluminium**)

R_i, R_e, R_{HC} – described in Figure 4.5 [m^2K/W]

$$R_{e,eps12.5} = R_{e,0,eps} - 2 \cdot \Delta R_{e,eps} = 1.00 - 2 \cdot 0.0569 = 0.886 \text{ (for } \mathbf{12.5\ cm}\text{)}$$

$$R_{e,xps15} = R_{e,0,xps} - 2 \cdot \Delta R_{e,xps15} = 1.35 - 2 \cdot 0.0502 = 1.25 \text{ (for } \mathbf{15\ cm}\text{)}$$

$$R_{e,xps20} = R_{e,0,xps} - 2 \cdot \Delta R_{e,xps20} = 1.35 - 2 \cdot 0.0377 = 1.275 \text{ (for } \mathbf{20\ cm}\text{)}$$

Due to small differences in this thermal resistance in extruded polystyrene (XPS) insulation, the average value was assumed: $\mathbf{R_{e,xps,\ with\ lamellas} = 1.26}$**, and** $\mathbf{R_{e,eps,\ with\ lamellas} = 0.886}$**.**

$$\Delta R_{e,eps} = \frac{d}{\lambda_{ins}} \cdot \frac{b}{T} = \frac{0.016}{0.036} \cdot \frac{0.016}{0.125} = 0.444 \cdot 0.128$$

$$= 0.0569 \text{ (for a pipe spacing of 12.5 cm)}$$

$$\Delta R_{e,xps15} = \frac{d}{\lambda_{ins}} \cdot \frac{b}{T} = \frac{0.016}{0.034} \cdot \frac{0.016}{0.15} = 0.471 \cdot 0.107$$

$$= 0.0502 \text{ (for a pipe spacing of 15 cm)}$$

$$\mathbf{\Delta R_{e,xps20} = 0.0377 \text{ (for pipe spacing 20 cm)}}$$

where

d – groove depth [m]
b – groove width [m]
λ_{ins} – thermal conductivity coefficient of EPS or XPS insulation [W/m·K]
$R_{se} = 0.17$ – heat transfer resistance under the floor surface [$m^2 \cdot K/W$]

$$\mathbf{R_{e,0,xps}} = R_u + R_{se} = \frac{d_k}{\lambda_k} = R_{se} = \frac{0.04}{0.034} + 0.17 = 1.176 + 0.17 = \mathbf{1.35} \text{ (}\mathbf{for\ XPS}\text{)}$$

$$\mathbf{R_{e,0,eps}} = R_u + R_{se} = \frac{d_k}{\lambda_k} = R_{se} = \frac{0.03}{0.036} + 0.17 = 0.833 + 0.17 = \mathbf{1.00} \text{ (}\mathbf{for\ XPS}\text{)}$$

where

d_k – thickness of EPS or XPS thermal insulation [m]

λ_k – thermal conductivity coefficient of thermal insulation [W/m·K]

$$\mathbf{R}_{\mathbf{U\,xps,\,eps}} = \frac{\mathbf{d_k}}{\boldsymbol{\lambda}_\mathbf{k}}$$

$$R_{U\,xps} = \frac{0.04}{0.034} = 1.176 \text{ (for XPS boards)}$$

$$R_{U\,eps} = \frac{0.03}{0.036} = 0.833 \text{ (for EPS boards)}$$

$$R_i = R_o + R_{con,i} + R_{si} = 0.0165 + 0.10 + 0.093 = 0.21$$

$$R_{con,i} = 0.10$$

Thermal resistance for polyurethane + ceramic tile adhesive

$$R_o = \frac{d_i}{\lambda_i} = \frac{0.002}{0.25} + \frac{0.009}{1.05} = 0.008 + 0.0085 = 0.0165$$

d_i – thickness of floor layers (adhesive and terracotta) [m]

λ_i – conductivity coefficient of floor layers (adhesive = 0.25 and terracotta = 1.05) [W/m·K]

Heat transfer resistance on the ceramic tile surface (assumed as in the case of most underfloor heating systems, according to the standard's suggestions [1])

$$R_{si} = \frac{1}{\alpha_i} = 0.093$$

T = 0.125; 0.15; 0.20 m – spacing of heating pipes

Ultimately, the maximum heat flux density depending on the heating coil spacing q_i^{max} is

The central zone:

$$q_{12,5}^{max} = 8.92 \cdot (7.164)^{1.1} = 8.92 \cdot 8.723 = \mathbf{77.81}\ \text{W/m}^2 \text{(for a pipe spacing of } \mathbf{12.5\ cm}\text{)}$$

$$q_{15}^{max} = 8.92 \cdot (6.615)^{1.1} = 8.92 \cdot 7.991 = \mathbf{71.28}\ \text{W/m}^2 \text{(for a pipe spacing of } \mathbf{15\ cm}\text{)}$$

$$q_{20}^{max} = 8.92 \cdot (5.463)^{1.1} = 8.92 \cdot 6.474 = \mathbf{57.75}\ \text{W/m}^2 \text{(for a pipe spacing of } \mathbf{20\ cm}\text{)}$$

The edge zone:

$$q_{12,5}^{max} = 8.92 \cdot (11.94)^{1.1} = 8.92 \cdot 15.30 = \mathbf{136.48}\ \text{W/m}^2 \text{(for a pipe spacing of } \mathbf{12.5\ cm}\text{)}$$

$$q_{15}^{max} = 8.92 \cdot (11.03)^{1.1} = 8.92 \cdot 14.02 = \mathbf{125.06}\ \text{W/m}^2 \text{(for a pipe spacing of } \mathbf{15\ cm}\text{)}$$

$$q_{20}^{max} = 8.92 \cdot (9.11)^{1.1} = 8.92 \cdot 11.36 = \mathbf{101.33}\ \text{W/m}^2 \text{(for a pipe spacing of } \mathbf{20\ cm}\text{)}$$

Substituting the above-calculated values, we calculate the balanced heat transfer coefficient K_H according to formula (4.2):

$$K_H = \frac{q_i^{max}}{\theta_F^{max} - \theta_i}$$

where

$q_{12,5}^{max} = \mathbf{77.81}$ W/ m^2(for a pipe spacing of **12.5 cm**)

$q_{15}^{max} = \mathbf{71.28}$ W/ m^2 (for a pipe spacing of **15 cm**)

$q_{20}^{max} = \mathbf{57.75}$ W/ m^2 (for a pipe spacing of **20 cm**)

$$K_H(12.5) = \frac{77.81}{29 - 20} = 8.646 \text{ W/m}^2\text{K}$$

$$K_H(15) = \frac{71.28}{29 - 20} = 7.92 \text{ W/m}^2\text{K}$$

$$K_H(20) = \frac{57.75}{29 - 20} = 6.417 \text{ W/m}^2\text{K}$$

Finally, we determine the computational density of the heat flux q_{ol}, emitted from the tested surface, using formula (4.1), taking into account the spacing used and the temperature feeding the surface heater with heat-dissipating lamellas. The same temperatures were used for calculations as those determined in the experimental tests:

$$q_{ol} = K_H : \text{times } \Delta\theta_{\mathbf{H}}$$

$$\Delta\theta_{\mathbf{H}} = \theta_{\mathbf{H}} - \theta_i$$

where

$\theta_{\mathbf{H}}$ – heating coil supply temperature [°C]

K_H – equivalent heat transmission coefficient [W/m^2·K]

θ_i – room temperature [°C]

q_{ol} – computational heat flux density (thermal power) on the floor surface with lamellas, depending on the heating coil spacing [W/m^2]

a. with a heating coil spacing of 12.5 cm, we have the following thermal powers q_{ol}:

$$q_{12.5}(28°\text{C}) = 8.646 \cdot (28 - 20) = 69.17 \text{ W/m}^2$$

$$q_{12.5}(35°\text{C}) = 8.646 \cdot (35 - 20) = 129.69 \text{ W/m}^2$$

$$q_{12.5}(40°\text{C}) = 8.646 \cdot (40 - 20) = 172.92 \text{ W/m}^2$$

$$q_{12.5}(45°\text{C}) = 8.646 \cdot (45 - 20) = 216.15 \text{ W/m}^2$$

b. with a heating coil spacing of 15 cm, we have the following thermal powers q_{ol}:

$$q_{15}(28°C) = 7.92 \cdot (28 - 20) = 63.36 \text{ W/m}^2$$

$$q_{15}(35°C) = 7.92 \cdot (35 - 20) = 118.80 \text{ W/m}^2$$

$$q_{15}(40°C) = 7.92 \cdot (40 - 20) = 158.40 \text{ W/m}^2$$

$$q_{15}(45°C) = 7.92 \cdot (45 - 20) = 198.00 \text{ W/m}^2$$

c. with a heating coil spacing of 20 cm, we have the following thermal powers q_{ol}:

$$q_{20}(28°C) = 6.417 \cdot (28 - 20) = 51.34 \text{ W/m}^2$$

$$q_{20}(35°C) = 6.417 \cdot (35 - 20) = 96.26 \text{ W/m}^2$$

$$q_{20}(40°C) = 6.417 \cdot (40 - 20) = 128.34 \text{ W/m}^2$$

$$q_{20}(45°C) = 6.417 \cdot (45 - 20) = 160.43 \text{ W/m}^2$$

4.2.2 Experimental Method for Calculating Heat Flux Density with Lamellas

Knowing the average temperature of the ceramic tile surface, ϑ_F (averaged from a thermal imaging camera), and the air temperature, the experimental heat flux density q_{el} emitted by the tested surface radiator was calculated using the formulas below. The formulas are based on the assumptions of the standard [1], considering the theoretically calculated coefficient of thermal conductivity of the dissipation layer (lamellas) from Section 4.2.1, which depends on the heating coil spacing. Instead of the maximum computed temperature values, those resulting from experimental tests were adopted, i.e.,

$$q_{el} = 8.92[k_{CL}(\vartheta_F - \vartheta_i)]^{1.1} \text{ [W/ m}^2] \tag{4.7}$$

$$q_i^{max} = 8.92\left(\overline{\theta}_F^{max} - \theta_i\right)^{1.1} \rightarrow q_i = 8.92(\overline{\vartheta}_F - \vartheta_i)^{1.1}$$

$$\overline{\theta}_F^{max} - \theta_i = k_{CL}\left(\theta_F^{max} - \theta_i\right) \rightarrow (\overline{\vartheta}_F - \vartheta_i) = k_{CL}(\vartheta_F - \vartheta_i)$$

q_{el} – experimental heat flux density of the tested radiator with lamellas, depending on heating coil spacing [W/m²]

$\overline{\vartheta}_F$ – average permissible temperature of the floor with lamellas
ϑ_F – average measurement temperature of the ceramic tile surface [°C]
ϑ_i – air temperature [°C]
k_{CL} – thermal conductivity coefficient of the dissipative layer (lamellas)
$k_{CL\,(12.5)} = 0.796$, $k_{CL\,(15)} = 0.735$, $k_{CL\,(20)} = 0.607$ (the results obtained in Section 4.2.1)

a. with a heating coil spacing of 12.5 cm, we have the following thermal powers q_{el}:

$$q_{12.5}(28^\circ C) = 8.92[0.796(28-20)]^{1.1} = 68.35\ W/m^2$$

$$q_{12.5}(35^\circ C) = 8.92[0.796(35-20)]^{1.1} = 136.48\ W/m^2$$

$$q_{12.5}(40^\circ C) = 8.92[0.796(40-20)]^{1.1} = 187.32\ W/m^2$$

$$q_{12.5}(45^\circ C) = 8.92[0.796(43.1-20)]^{1.1} = 219.52\ W/m^2$$

b. with a heating coil spacing of 15 cm, we have the following thermal powers q_{el}:

$$q_{15}(28^\circ C) = 8.92[0.735(28-20)]^{1.1} = 62.62\ W/m^2$$

$$q_{15}(35^\circ C) = 8.92[0.735(34.9-20)]^{1.1} = 124.08\ W/m^2$$

$$q_{15}(40^\circ C) = 8.92[0.735(38.7-20)]^{1.1} = 159.40\ W/m^2$$

$$q_{15}(45^\circ C) = 8.92[0.735(40.6-20)]^{1.1} = 177.24\ W/m^2$$

c. with a heating coil spacing of 20 cm, we have the following thermal powers q_{el}:

$$q_{20}(28^\circ C) = 8.92[0.607(27.2-20)]^{1.1} = 45.14\ W/m^2$$

$$q_{20}(35^\circ C) = 8.92[0.607(31.8-20)]^{1.1} = 77.78\ W/m^2$$

$$q_{20}(40^\circ C) = 8.92[0.607(33.2-20)]^{1.1} = 87.95\ W/m^2$$

$$q_{20}(45^\circ C) = 8.92[0.607(34.3-20)]^{1.1} = 96.07\ W/m^2$$

The relationships between the heat flux density q_{el} and the supply temperature ϑ_V for radiators with lamellas, calculated from the formulas, and based on measurements with a thermal imaging camera of the radiators' average surface temperature, are also shown in Figure 4.6. The three lower dashed lines (their value depends on the heating coil spacing), calculated q_i^{max} from formula (4.3) in Figure 4.6, show the limit values of the central zone's heat flux and the edge zone's upper ones. Exceeding these values is undesirable due to the effect on the comfort of the room humans use. Their value depends on the spacing of the heating coils.

Similar data from Figure 4.6 is illustrated by a graph of the dependence of thermal efficiency qF on the supply temperature θ_V, with dissipative lamellas according to [7]. This is shown in Figure 4.7, without taking into consideration temperature losses along the length $\delta T = 0°C$, and with their assumption at $\delta T = 5°C$.

Table 4.1 compares the thermal efficiency data q_{el} from Figure 4.6 to the q_F data from Figure 4.7.

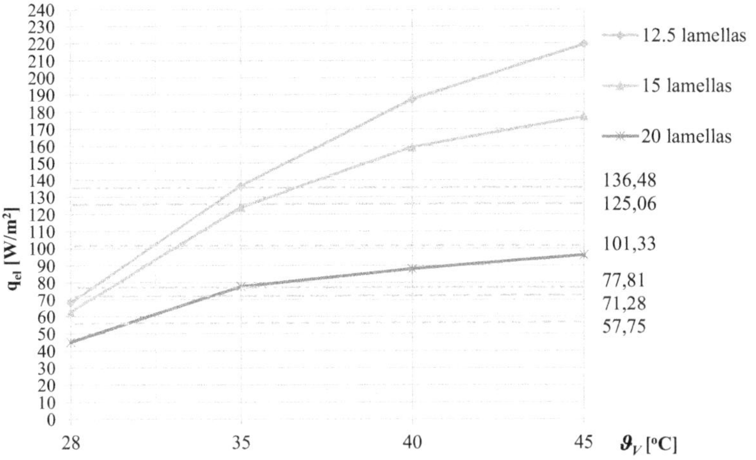

FIGURE 4.6 Dependence of the q_{el} value on the supply temperature ϑ_V for heating models with lamellas (Source: Prepared by Karpiesiuk J.), where – – – – – – – – – – q limit for a heating coil spacing of 20 cm, — — — — — — q limit for a heating coil spacing of 15 cm, —··— ··—··—·q limit for a heating coil spacing of 12.5 cm.

TABLE 4.1
Experimental Value of Thermal Efficiency of Lightweight Heated Floor *with Lamellas* and Company's Catalogue [7]

Heating Coil Spacing [cm]	Computational Thermal Efficiency q_{el} [W/m²]				Thermal Efficiency from the Catalogue q_F [W/m²]			
	Lightweight Floor Radiator with Lamellas							
	Heating Coil Supply Temperature [°C]							
	28	**35**	**40**	**45**	**28**	**35**	**40**	**45**
12.5	68.35	136.48	187.32	219.52	69.2	129.7	172.9	216.2
15	62.62	124.08	159.4	177.24	63.4	118.8	158.4	198.0
	$q_{12.5}^{max} = \mathbf{77.81}, q_{15}^{max} = \mathbf{71.28}, q_{12.5}^{max}$							

4.2.3 The Maximum Temperature of the Medium in the Heating Pipe Using the NORDTEST NT VVS 127 Method with Lamellas

Calculations of the maximum allowable temperature of the medium in the heating pipe θ_H^{max} were made according to formula (4.5) for assumed distances:

$$\theta_H^{max} = \theta_i + q_i^{max}\left(R_i + \frac{1}{\eta} \cdot R_{HC}\right)$$

$$\theta_H^{max},(12.5) = 20 + 77.81(0.21 + 1.24 \cdot 0.059) = 42.0°C$$

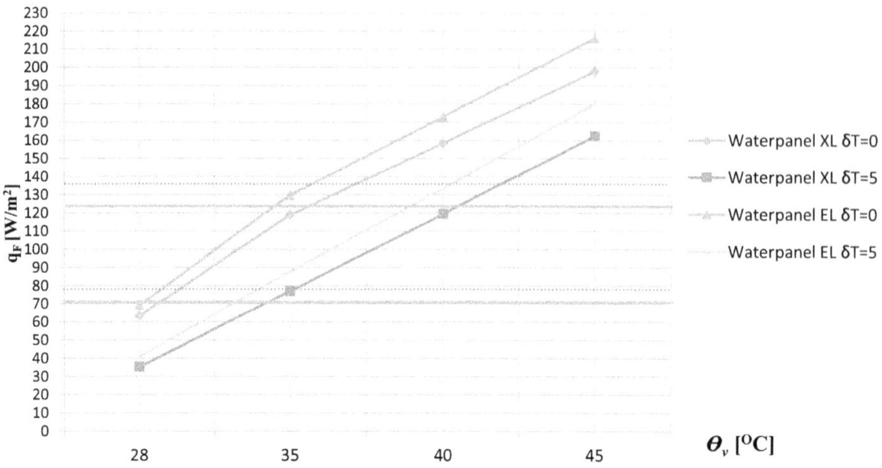

FIGURE 4.7 Dependence of efficiency q_F on the supply temperature θ_V, with the difference in supply and return temperatures δT for heating stations *with fins* (Waterpanel EL with a coil spacing of 12.5 mm, Waterpanel XL with a coil spacing of 15 cm), on a ceramic tile floor given in the catalogue [7] (Source: Prepared by Karpiesiuk J.), where ----- q lower border (in the central zone) and upper border (in the edge zone) with a heating coil spacing of 15 cm ——— q lower border (in the central zone) and upper border (in the edge zone) with a heating coil spacing of 12.5 cm.

$$\theta_{\mathbf{H}}^{\mathbf{max}},(15) = 20 + 71.28(0.21 + 1.17 \cdot 0.084) = 42.0°\mathrm{C}$$

$$\theta_{\mathbf{H}}^{\mathbf{max}},(20) = 20 + 57.75(0.21 + 1.17 \cdot 0.144) = 41.9°\mathrm{C}$$

The following are the calculations of the individual components of the formula (4.5) for the maximum permissible temperature of the medium in the heating pipe $\theta_{\mathbf{H}}^{\mathbf{max}}$, with appropriate heating coil spacing, leading to its determination.

The efficiency of the underfloor heating system with a heating coil spacing of 12.5 cm is

$$\eta_{12.5} = \frac{R_e}{R_i + R_e} = \frac{0.886}{0.21 + 0.886} = \mathbf{0.808}$$

The efficiency of the underfloor heating system with heating coil spacings of 15 and 20 cm is

$$\eta_{15 \text{ and } 20} = \frac{R_e}{R_i + R_e} = \frac{1.26}{0.21 + 1.26} = 0.857$$

$$R_{HC} = T \cdot R_R' + T \cdot R_{R,con}' \frac{T}{2} \cdot R_U' + R_{CL}$$

where

$$R_{HC,12.5} = T \cdot R'_R + T \cdot R'_{R,con} \frac{T}{2} \cdot R'_U + R_{CL,12.5}$$
$$= 0.125 \cdot 0.107 + 0.125 \cdot 0.625 \cdot 0.0625 \cdot 0.5 + 0.044$$
$$= 0.013 + 0.002 + 0.044 = \mathbf{0.059}$$

$$R_{HC,15} = T \cdot R'_R + T \cdot R'_{R,con} \frac{T}{2} \cdot R'_U + R_{CL,15} = 0.15 \cdot 0.107 + 0.15 \cdot 0.625 \cdot 0.075 \cdot 0.5 + 0.065$$
$$= 0.016 + 0.003 + 0.065 = \mathbf{0.084}$$

$$R_{HC,20} = T \cdot R'_R + T \cdot R'_{R,con} \frac{T}{2} \cdot R'_U + R_{CL,20} = 0.20 \cdot 0.107 + 0.20 \cdot 0.625 \cdot 0.1 \cdot 0.5 + 0.117$$
$$= 0.021 + 0.006 + 0.117 = \mathbf{0.144}$$

$$R'_U = \frac{0.008}{d_a} = 0.50$$

$$R'_{R,con} = \frac{0.01}{d_a} = 0.625$$

$$R'_R = \frac{1}{2\pi\lambda_R} \ln\left(\frac{d_o}{d_{ip}}\right) = 0.107$$

where
$d_a = 0.016$ [m]
$d_o = 0.016$ [m]
$d_{ip} = 0.012$ [m]
$\lambda_R = 0.43$ [W/m°C]

$\ln\left(\frac{0.016}{0.012}\right) = \ln(1.33) = 0.29$ according to http://www.calculat.org/pl/logarytmy/logarytm-naturalny.html

$$R_{CL,12.5} = \frac{1}{\frac{1}{R_i} + \frac{1}{R_e}}\left(\frac{1}{k_{CL}} - 1\right) = \frac{1}{\frac{1}{0.21} + \frac{1}{0.886}}\left(\frac{1}{0.796} - 1\right)$$
$$= \frac{1}{4.76 + 1.13}(1.26 - 1) = 0.044$$

$$R_{CL,15} = \frac{1}{\frac{1}{0.21} + \frac{1}{1.26}}\left(\frac{1}{0.735} - 1\right) = \frac{1}{4.76 + 0.79}(1.36 - 1) = 0.065$$

$$R_{CL,20} = \frac{1}{\frac{1}{0.21}+\frac{1}{1.26}}\left(\frac{1}{0.607}-1\right) = \frac{1}{4.76+0.79}(1.65-1) = 0.117$$

$$R_{i,\text{ with lamella}} = 0.21$$

$$R_{e,xps,\text{ with lamellas}} = 1.26$$

$$R_{e,eps,\text{ with lamellas}} = 0.886$$

Based on the above calculations, it can be assumed that the permissible temperature of the medium $\boldsymbol{\theta_H^{max}}$ in the heating pipe (16×2mm, with a heating coil spacing of 12.5–20cm, where the use of heat-conducting lamellas 0.09mm thick is placed on EPS and XPS thermal insulation 30–40mm thick) is **42°C**.

4.3 METHODOLOGY FOR CALCULATING THE THERMAL POWER OF A LIGHTWEIGHT RADIANT HEATING SYSTEM WITHOUT LAMELLAS

So far, no formulas or procedures have been developed for calculating the thermal power of a lightweight underfloor heating system without lamellas. Designers and scientists still use existing standards (e.g., PN-EN 15377-1:2008) [8], which are not adapted for this purpose, and the NT VVS127:2001 [1] standard, which can only be used to calculate lightweight systems with only a conductive layer. This chapter presents the calculation of the thermal power of LRHS using an original method based on the NT VVS 127:2001 standard [1], but without heat-conducting lamellas. The author's proposed procedure for calculating the maximum LRHS heat flux density without a dissipative layer (lamellas) $\mathbf{q_{ob}}$ is based on the formulas presented below. To determine the design thermal power q, the basic formulas marked as (4.8), (4.9), (4.10), and (4.11), the auxiliary formulas presented below (without numerical designations) should be used to determine these four main formulas:

$$q_{ob} = K_{Hb} \cdot \Delta\theta_H \text{ [W/m}^2\text{]} \tag{4.8}$$

where

q_{ob} – computational heat flux density (thermal power) on the floor surface without lamellas [W/m²]

$\Delta\theta_H = \theta_H - \theta_i$

θ_H – heating coil supply temperature [°C]

θ_i – room temperature [°C]

$$K_{Hb} = \frac{q_B^{max}}{\theta_F^{max} - \theta_i} \text{ [W/ m}^2\text{K]} \tag{4.9}$$

where

K_{Hb} – equivalent heat transmission coefficient without lamellas [W/m²·K]

θ_F^{max} – maximum permissible floor surface temperature [°C] (in the middle zone equal to 29°C)

q_B^{max} – maximum heat flux density (without lamellas) [W/m²]

θ_i – room temperature [°C]

$$q_B^{max} = 8.92(\bar{\theta}_F^{max} - \theta_i)^{1.1}\ [W/m^2] \tag{4.10}$$

where

q_B^{max} – maximum heat flux density, depending on the coil spacing without lamellas [W/m²]

$\bar{\theta}_F^{max}$ – θ is calculated from the formula:

$\bar{\theta}_F^{max} - \theta_i = k_{CB}(\theta_F^{max} - \theta_i)\ [°C]$, where

θ_F^{max} – maximum permissible floor surface temperature, equal to 29°C in the central zone.

k_{CB} – thermal conductivity coefficient without dissipation layer (without lamellas) [W/m²·K]

θ_i – room temperature [°C]

k_{CB} – coefficient is calculated according to the following formulas:

$$k_{CB} = \frac{L_U + 2L_k \cdot k_k}{T}\ [W/m^2K] \tag{4.11}$$

L_U – external diameter of the heating pipe [m]

T – spacing of heating pipes (coils) [m]

$$L_k = \frac{T - L_U}{2}$$

L_k – estimated width of the horizontal heat-conducting adhesive layer and a ceramic tile floor [m]

$$k_k = \frac{\ell}{L_k}\tanh\left(\frac{L_k}{\ell}\right)$$

k_k – thermal conductivity coefficient of horizontal layers above thermal insulation

$$\ell_B = \sqrt{\frac{S_{WB} \cdot \lambda_{WB}}{\frac{1}{R_i} + \frac{1}{R_e}}}$$

ℓ_B – characteristic length of the adhesive layer with the floor [m]

S_{WB} – thickness of the conductive layer (adhesive + floor) [m]

λ_{WB} – average thermal conductivity weight value of the adhesive layer and floor, with all their thickness [W/m·K]

R_i – thermal resistance of the layer conducting heat (floor adhesive) to the heated room [m^2 K/W]

R_e – thermal resistance under the layer that conducts heat (floor adhesive) to the room below or outside [m^2 K/W]

$$R_e = R_{e,0} - 2 \cdot \Delta R_e$$

$$\Delta R_e = \frac{d}{\lambda_{ins}} \cdot \frac{b}{T}$$

d – groove depth [m]

b – groove width [m]

λ_{ins} – thermal conductivity coefficient of thermal insulation [W/m·K]

T – spacing of heating pipes (coils) [m]

$$R_{e,0} = R_u + R_{se} = \frac{d_k}{\lambda_k} + R_{se}$$

R_{se} – heat transfer resistance under the floor surface [m^2 K/W]

R_u – thermal insulation resistance [m^2 K/W]

d_k – thermal insulation thickness [m]

λ_k – thermal conductivity coefficient of thermal insulation [W/m·K]

$$R_i = R_o + R_{con,i} + R_{si}$$

$R_{con,i}$ – resistance for conductive layers carefully shaped and permanently glued, with a layer of insulation [m^2·K/W]

$R_o = \frac{d_i}{\lambda_i}$ – thermal resistance of floor layers above the conductive layer of floor adhesive [m^2 K/W]

d_i – thickness of individual floor layers above the conductive layer [m]

λ_i – thermal conductivity coefficient of individual floor layers above the conductive **layer** [W/(m·K)]

$R_{si} = \frac{1}{\alpha_i}$ – heat transfer resistance on the floor surface [m^2 K/W]

α_i – heat transfer coefficient on the floor surface [W/m^2·K]

4.3.1 Theoretical Method for Calculating the Heat Flux Density, Using the NORDTEST NT VVS 127 Method, without Lamellas

This subchapter provides an example of determining the design thermal efficiency (heat flux density) based on the data, materials, and layers of a lightweight floor system with ceramic tile flooring, without lamellas as described in Section 4.1, using the formulas from Section 4.3. Here are the theoretical calculations of the maximum heat flux density in the floor without lamellas, depending on the distance between the heating coils, using the author's own formulas (4.8)–(4.11), based on the NT VVS127:2001 standard [1]:

$$q_B^{max} = 8.92(\bar{\theta}_F^{max} - \theta_i)^{1.1}$$

where

$$\bar{\theta}_F^{max} - \theta_i = k_{CB}(\theta_F^{max} - \theta_i)\,[^\circ C]$$

$$k_{CB(12.5)} = \frac{L_U + 2L_k \cdot k_k}{T} = \frac{0.016 + 2 \cdot 0.0545 \cdot 0.722}{0.125} = 0.758$$

$$k_{CB(15)} = \frac{L_U + 2L_k \cdot k_k}{T} = \frac{0.016 + 2 \cdot 0.067 \cdot 0.655}{0.15} = 0.692$$

$$k_{CB(20)} = \frac{L_U + 2L_k \cdot k_k}{T} = \frac{0.016 + 2 \cdot 0.092 \cdot 0.521}{0.20} = 0.559$$

Substituting the values *of* k_{CB} with specific heating coil spacing, we have the following formula:

In the middle zone:

$$\bar{\theta}_F^{max} - \theta_i = k_{CB}(\theta_F^{max} - \theta_i)$$

$$0.758(29 - 20) = 6.822^\circ C \quad \text{(for a pipe spacing of } \mathbf{12.5\ cm}\text{)}$$

$$0.692(29 - 20) = 6.228^\circ C \quad \text{(for a pipe spacing of } \mathbf{15\ cm}\text{)}$$

$$0.559(29 - 20) = 5.034^\circ C \quad \text{(for a pipe spacing of } \mathbf{20\ cm}\text{)}$$

In the edge zone:

$$\bar{\theta}_F^{max} - \theta_i = k_{CB}(\theta_F^{max} - \theta_i)$$

$$0.758(35 - 20) = 11.37^\circ C \quad \text{(for a pipe spacing of } \mathbf{12.5\ cm}\text{)}$$

$$0.692(35 - 20) = 10.38^\circ C \quad \text{(for a pipe spacing of } \mathbf{15\ cm}\text{)}$$

$$0.559(35 - 20) = 8.39^\circ C \quad \text{(for a pipe spacing of } \mathbf{20\ cm}\text{)}$$

The formula leading to the determination of the thermal conductivity coefficient of the adhesive layer and the flooring without lamellas k_{CB}, with appropriate heating coil spacing, assuming no gap in this layer:

$L_U = 0.016\,[m]$ – external diameter of the heating pipe

$$L_k = \frac{T - L_U}{2} = \frac{0.125 - 0.016}{2}$$

$$L_k = 0.0545\ m \quad \text{(for a pipe spacing of } \mathbf{12.5\ cm}\text{)}$$

$$L_k = 0.067\ m \quad \text{(for a pipe spacing of } \mathbf{15\ cm}\text{)}$$

$$L_k = 0.092\ m \quad \text{(for a pipe spacing of } \mathbf{20\ cm}\text{)}$$

$$k_k = \frac{\ell}{L_k} \tanh\left(\frac{L_k}{\ell}\right)$$

$$k_k = \frac{0.0488}{0.0545} \tanh 1.1168 = 0.895 \cdot 0.806 = 0.722 \quad \text{(for a pipe spacing of } \mathbf{12.5\ cm}\text{)}$$

$$k_k = \frac{0.0505}{0.067} \tanh 1.3267 = 0.754 \cdot 0.868 = 0.655 \quad \text{(for a pipe spacing of } \mathbf{15\ cm}\text{)}$$

$$k_k = \frac{0.0505}{0.092} \tanh 1.8218 = 0.549 \cdot 0.949 = 0.521 \quad \text{(for a pipe spacing of } \mathbf{20\ cm}\text{)}$$

Calculations of hyperbolic tangent according to http://calcoolator.pl/tangens_kalkulator.html

$$\ell_{B\ xps} = \sqrt{\frac{\mathbf{S_{WB} \cdot \lambda_{WB}}}{\frac{\mathbf{1}}{\mathbf{R_i}} + \frac{\mathbf{1}}{\mathbf{R_e}}}} = \sqrt{\frac{\mathbf{0.012 \cdot 1}}{\frac{\mathbf{1}}{\mathbf{0.255}} + \frac{\mathbf{1}}{\mathbf{1.26}}}} = \sqrt{\frac{\mathbf{0.012}}{\mathbf{3.92 + 0.79}}}$$

$$= 0.0505 \left(\text{for pipe spacings of 15 and 20 cm}\right)$$

$$\ell_{B\ eps} = \sqrt{\frac{\mathbf{S_{WB} \cdot \lambda_{WB}}}{\frac{\mathbf{1}}{\mathbf{R_i}} + \frac{\mathbf{1}}{\mathbf{R_e}}}} = \sqrt{\frac{\mathbf{0.012 \cdot 1}}{\frac{\mathbf{1}}{\mathbf{0.255}} + \frac{\mathbf{1}}{\mathbf{0.886}}}} = \sqrt{\frac{\mathbf{0.012}}{\mathbf{3.92 + 1.13}}}$$

$$= 0.0488 \left(\text{for a pipe spacing of 12.5 cm}\right)$$

where

$S_{WB} = 0.012$ m – thickness of the conductive layer (adhesive + ceramic tile)

$\lambda_{WB} = 1.0$ W/mK – the average weight value of thermal conductivity of the adhesive layer and ceramic tile flooring, taking into account their thickness:

$$\lambda_{WB} = \frac{3\ \text{mm}}{12\ \text{mm}} 0.85 + \frac{9\ \text{mm}}{12\ \text{mm}} 1.05 = 0.21 + 0.79 = 1.0$$

where

3 mm – adhesive thickness; 9 mm – tile thickness

R_i, R_e, – described in Figure 4.5 [m^2 K/W]

$$R_{e,eps12.5} = R_{e,0,eps} - 2 \cdot \Delta R_{e,eps} = 1.00 - 2 \cdot 0.0569 = 0.886 \text{ (a spacing of } \mathbf{12.5\ cm}\text{)}$$

$$R_{e,xps15} = R_{e,0,xps} - 2 \cdot \Delta R_{e,xps15} = 1.35 - 2 \cdot 0.0502 = 1.25 \text{ (a spacing of } \mathbf{15\ cm}\text{)}$$

$$R_{e,xps20} = R_{e,0,xps} - 2 \cdot \Delta R_{e,xps20} = 1.35 - 2 \cdot 0.0377 = 1.275 \text{ (a spacing of } \mathbf{20\ cm}\text{)}$$

Due to the small differences in this thermal resistance at XPS, the average value was assumed: **$R_{e,\text{xps, without lamellas}} = 1.26$, and $R_{e,\text{eps, without lamellas}} = 0.886$.**

$$\Delta R_{e,eps} = \frac{d}{\lambda_{ins}} \cdot \frac{b}{T} = \frac{0.016}{0.036} \cdot \frac{0.016}{0.125} = 0.444 \cdot 0.128$$

$$= 0.0569 \text{ (for a pipe spacing of } \mathbf{12.5\ cm}\text{)}$$

$$\Delta R_{e,xps15} = \frac{d}{\lambda_{ins}} \cdot \frac{b}{T} = \frac{0.016}{0.034} \cdot \frac{0.016}{0.15} = 0.471 \cdot 0.107$$

$$= 0.0502 \text{ (for a pipe spacing of } \mathbf{15\ cm}\text{)}$$

$$\Delta R_{e,xps20} = 0.0377 \text{ (for a pipe spacing of } \mathbf{20\ cm}\text{)}$$

d – groove depth [m],
b – groove width [m],
λ_{ins} – conductivity coefficient thermal insulation XPS or EPS [W/m K]
R_{se} = 0.17 – heat transfer resistance under the floor surface [m²·K/W]

$$\mathbf{R_{e,0,xps}} = R_u + R_{se} = \frac{d_k}{\lambda_k} + R_{se} = \frac{0.04}{0.034} + 0.17 = 1.176 + 0.17 = \mathbf{1.35} \text{ } \mathbf{(for\ XPS)}$$

$$\mathbf{R_{e,0,eps}} = R_u + R_{se} = \frac{d_k}{\lambda_k} + R_{se} = \frac{0.03}{0.036} + 0.17 = 0.833 + 0.17 = \mathbf{1.00} \text{ } \mathbf{(for\ EPS)}$$

d_k – thermal insulation thickness [m]
λ_k – thermal conductivity coefficient of thermal insulation [W/mK]

$$R_u = \frac{1}{U_{xps}} = \frac{0.04}{0.034} = 1.176 \text{ m}^2\text{K/ W (for XPS boards)}$$

$$R_u = \frac{1}{U_{eps}} = \frac{0.03}{0.036} = 0.833 \text{ m}^2\text{K/ W (for EPS boards)}$$

R_u – thermal resistance of EPS or XPS thermal insulation

$$R_i = R_o + R_{con,i} + R_{si} = 0.012 + 0.15 + 0.093 = 0.255 \text{ m}^2{}^\circ\text{C/W}$$

$R_{con,i} = 0.15\ m^{2\circ}C/W$ – resistance for conductive layers (adhesive and ceramic tile), assuming the direct contact of these layers

$R_o = \frac{d_i}{\lambda_i} = \frac{0.003}{0.85} + \frac{0.009}{1.05} = 0.0035 + 0.0085 = 0.012\ m^2\,{}^\circ C/W$ (thermal resistance for adhesive mortar + ceramic tile)

$R_{si} = \frac{1}{\alpha_i} = 0.093\ m^2\,{}^\circ C/W$ – heat transfer resistance on the tile surface (assumed, as in the case of most underfloor heating systems, according to the suggestions of the standard [1])

T = 0.125; 0.15; 0.20 m – spacing of heating pipes

Ultimately, the maximum heat flux density, depending on the heating coil spacing q_B^{max} without lamellas, is

The central zone:

$$q_{12.5}^{max} = 8.92 \cdot (6.822)^{1.1} = 8.92 \cdot 8.266 = \mathbf{73.73}\ \mathrm{W/m^2}\ \text{(for a pipe spacing of } \mathbf{12.5\ cm}\text{)}$$

$$q_{15}^{max} = 8.92 \cdot (6.228)^{1.1} = 8.92 \cdot 7.478 = \mathbf{66.70}\ \mathrm{W/m^2}\ \text{(for a pipe spacing of } \mathbf{15\ cm}\text{)}$$

$$q_{20}^{max} = 8.92 \cdot (5.034)^{1.1} = 8.92 \cdot 5.917 = \mathbf{52.78}\ \mathrm{W/m^2}\ \text{(for a pipe spacing of } \mathbf{20\ cm}\text{)}$$

The edge zone:

$$q_{12.5}^{max} = 8.92 \cdot (11.37)^{1.1} = 8.92 \cdot 14.50 = \mathbf{129.34}\ \mathrm{W/m^2}\ \text{(for a pipe spacing of } \mathbf{12.5\ cm}\text{)}$$

$$q_{15}^{max} = 8.92 \cdot (10.38)^{1.1} = 8.92 \cdot 13.09 = \mathbf{116.76}\ \mathrm{W/m^2}\ \text{(for a pipe spacing of } \mathbf{15\ cm}\text{)}$$

$$q_{20}^{max} = 8.92 \cdot (8.39)^{1.1} = 8.92 \cdot 10.38 = \mathbf{92.59}\ \mathrm{W/m^2}\ \text{(for a pipe spacing of } \mathbf{20\ cm}\text{)}$$

The computational density of the heat flux emitted from the tested surface, $\boldsymbol{q_{ob}}$, taking into account the spacing used and the temperature supplied to the surface heater without heat-dissipating lamellas, will be obtained using formula (4.8). The same temperatures were used for calculations as those determined in the experimental tests:

$$q_{ob} = K_{Hb} : \text{times}\ \Delta\theta_{\mathbf{H}}\ [\mathrm{W/m^2}]$$

$$\Delta\theta_{\mathbf{H}} = \theta_{\mathbf{H}} - \theta_{\mathbf{i}}$$

where

$\theta_{\mathbf{H}}$ – heating coil supply temperature [°C]

K_{Hb} – equivalent heat transmission coefficient without lamellas [W/m^2·K]

θ_i – room temperature [°C]

q_{ob} – computational heat flux density (thermal efficiency) on the floor surface without lamellas, depending on the heating coil spacing [W/m^2]

$$q_{12.5}^{max} = \mathbf{73.73}\ \mathrm{W/m^2}\ \left(\text{for a pipe spacing of } \mathbf{12.5\ cm}\right)$$

$$q_{15}^{max} = \mathbf{66.70}\ \mathrm{W/m^2}\ \left(\text{for a pipe spacing of } \mathbf{15\ cm}\right)$$

$$q_{20}^{max} = \mathbf{52.78}\ \mathrm{W/m^2}\ \left(\text{for a pipe spacing of } \mathbf{20\ cm}\right)$$

$$K_{Hb},(12.5) = \frac{73.73}{29 - 20} = 8.192\ \mathrm{W/m^2K}$$

$$K_{Hb},(15) = \frac{66.70}{29 - 20} = 7.411\ \mathrm{W/m^2K}$$

$$K_{Hb},(20) = \frac{52.78}{29 - 20} = 5.864\ \mathrm{W/m^2K}$$

a. With a heating coil spacing of 12.5 cm, we have the following thermal powers q_{ob}:

$$q_{12.5}(28°C) = 8.192 \cdot (28 - 20) = 65.54 \text{ W/m}^2$$

$$q_{12.5}(35°C) = 8.192 \cdot (35 - 20) = 122.88 \text{ W/m}^2$$

$$q_{12.5}(40°C) = 8.192 \cdot (40 - 20) = 163.84 \text{ W/m}^2$$

$$q_{12.5}(45°C) = 8.192 \cdot (45 - 20) = 204.80 \text{ W/m}^2$$

b. With a heating coil spacing of 15 cm, we have the following thermal powers q_{ob}:

$$q_{15}(28°C) = 7.411 \cdot (28 - 20) = 59.29 \text{ W/m}^2$$

$$q_{15}(35°C) = 7.411 \cdot (35 - 20) = 111.17 \text{ W/m}^2$$

$$q_{15}(40°C) = 7.411 \cdot (40 - 20) = 148.22 \text{ W/m}^2$$

$$q_{15}(45°C) = 7.411 \cdot (45 - 20) = 185.28 \text{ W/m}^2$$

c. With a heating coil spacing of 20 cm, we have the following thermal powers q_{ob}:

$$q_{20}(28°C) = 5.864 \cdot (28 - 20) = 46.91 \text{ W/m}^2$$

$$q_{20}(35°C) = 5.864 \cdot (35 - 20) = 87.97 \text{ W/m}^2$$

$$q_{20}(40°C) = 5.864 \cdot (40 - 20) = 117.28 \text{ W/m}^2$$

$$q_{20}(45°C) = 5.864 \cdot (45 - 20) = 146.60 \text{ W/m}^2$$

4.3.2 Experimental Method for Calculating Heat Flux Density without Lamellas

Knowing the average temperature of the ceramic tile surface, ϑ_F (averaged from a thermal imaging camera), and the air temperature ϑ_i, the experimental heat flux density q_{eb} emitted by the tested surface radiator was calculated. Two different formulas (4.12) and (4.13) were used. Formula (4.12) is based on the assumptions of the standard PN EN 15377-1:2008 [8]. Formula (4.13) was adopted, taking into account the theoretically calculated thermal conductivity coefficients k_{CB}, without the dissipation layer from Section 4.3.1, which depend on the coil spacing. Instead of the maximum calculated temperature values, the experimental test results were adopted.

4.3.2.1 Calculations of the Thermal Power of a Radiator without Lamellas Following the Standard [8]

$$q_e = 8.92(\vartheta_F - \vartheta_i)^{1.1} \text{ [W/ m}^2\text{]} \quad (4.12)$$

where

q_e – experimental heat flux density of the tested radiator without lamellas, depending on the heating coil spacing [W/m²]

ϑ_F – average temperature of the ceramic tile surface from the measurements [°C]

ϑ_i – air temperature [°C]

a. With a heating coil spacing of 12.5 cm, we have the following thermal powers q_e:

$$q_{12.5}(28°C) = 8.92(28 - 20)^{1.1} = 87.90\ W/m^2$$

$$q_{12.5}(35°C) = 8.92(34.1 - 20)^{1.1} = 163.86\ W/m^2$$

$$q_{12.5}(40°C) = 8.92(36.2 - 20)^{1.1} = 190.91\ W/m^2$$

$$q_{12.5}(45°C) = 8.92(38.3 - 20)^{1.1} = 218.30\ W/m^2$$

b. With a heating coil spacing of 15 cm, we have the following thermal powers q_e:

$$q_{15}(28°C) = 8.92(27.8 - 20)^{1.1} = 85.44\ W/\ m^2$$

$$q_{15}(35°C) = 8.92(33 - 20)^{1.1} = 149.87\ W/\ m^2$$

$$q_{15}(40°C) = 8.92(34.7 - 20)^{1.1} = 171.56\ W/\ m^2$$

$$q_{15}(45°C) = 8.92(36.8 - 20)^{1.1} = 198.70\ W/\ m^2$$

c. With a heating coil spacing of 20 cm, we have the following thermal powers q_e:

$$q_{20}(28°C) = 8.92(27.3 - 20)^{1.1} = 79.44\ W/\ m^2$$

$$q_{20}(35°C) = 8.92(29.6 - 20)^{1.1} = 107.40\ W/\ m^2$$

$$q_{20}(40°C) = 8.92(30.6 - 20)^{1.1} = 119.70\ W/\ m^2$$

$$q_{20}(45°C) = 8.92(31.7 - 20)^{1.1} = 133.50\ W/\ m^2$$

4.3.2.2 Calculations of the Thermal Power of a Radiator without Lamellas, Taking into Account the k_{CB} Coefficient

$$q_{eb} = 8.92[k_{CB}(\vartheta_F - \vartheta_i)]^{1.1}\ [W/\ m^2] \tag{4.13}$$

q_{eb} – experimental heat flux density of the tested radiator without lamellas (author's calculations) [W/m^2]

The author's own calculation procedure follows point 4.3, where the floor temperature data were assumed to be as in point A, obtained from experimental tests of a radiant heater without lamellas. Values of the thermal conductivity coefficient without the dissipation layer k_{CB} are calculated in Section 4.3.1. Its values, depending on the coil spacing, are as follows:

$$k_{CB(12.5)} = 0.758, k_{CB(15)} = 0.692, k_{CB(20)} = 0.559, \text{hence}$$

a. with a heating coil spacing of 12.5 cm, we have the following thermal powers q_{eb}:

$$q_{12.5}(28°C) = 8.92[0.758(28 - 20)]^{1.1} = 64.76 \text{ W/m}^2$$

$$q_{12.5}(35°C) = 8.92[0.758(34.1 - 20)]^{1.1} = 120.87 \text{ W/m}^2$$

$$q_{12.5}(40°C) = 8.92[0.758(36.2 - 20)]^{1.1} = 140.76 \text{ W/m}^2$$

$$q_{12.5}(45°C) = 8.92[0.758(38.3 - 20)]^{1.1} = 160.92 \text{ W/m}^2$$

b. with a heating coil spacing of 15 cm, we have the following thermal powers q_{eb}:

$$q_{15}(28°C) = 8.92[0.692(27.8 - 20)]^{1.1} = 57.00 \text{ W/m}^2$$

$$q_{15}(35°C) = 8.92[0.692(33 - 20)]^{1.1} = 100.00 \text{ W/m}^2$$

$$q_{15}(40°C) = 8.92[0.692(34.7 - 20)]^{1.1} = 114.44 \text{ W/m}^2$$

$$q_{15}(45°C) = 8.92[0.692(36.8 - 20)]^{1.1} = 132.55 \text{ W/m}^2$$

c. with a heating coil spacing of 20 cm, we have the following thermal powers q_{eb}:

$$q_{20}(28°C) = 8.92[0.559(27.3 - 20)]^{1.1} = 41.92 \text{ W/m}^2$$

$$q_{20}(35°C) = 8.92[0.559(29.6 - 20)]^{1.1} = 56.64 \text{ W/m}^2$$

$$q_{20}(40°C) = 8.92[0.559(30.6 - 20)]^{1.1} = 63.24 \text{ W/m}^2$$

$$q_{20}(45°C) = 8.92[0.559(31.7 - 20)]^{1.1} = 70.38 \text{ W/m}^2$$

The dependencies calculated from the heat flux density formulas qe on the supply temperature ϑ_V for radiators without lamellas, based on measurements with a thermal imaging camera of the average surface temperature of the radiators, are shown in Figure 4.8. Figure 4.8 shows the dependence of the heat flux density on the supply temperature of the heating coil on the floor without lamellas according to the standard PN-EN 15377-1:2008 [8]. The lower and upper dashed lines show the limiting lower (100 W/m^2) and upper (175.46 W/m^2) q values for temperatures, respectively, of 29°C and 35°C. The dashed-dot line shows the limit value of q at a temperature of 40°C, as the maximum for a surface wall radiator. Exceeding these values is undesirable due to the need for comfort in human room use.

A graph with divergent data on thermal power calculated from the PN-EN 15377-1:2008 standard [8], shown in graph 4.8 without the use of heat-dissipating lamellas, can be found in [7]. This is shown in Figure 4.9, without taking into account temperature losses along the length δT = 0°C and assuming them at δT = 5°C. The data from Figure 4.9 are similar to the data calculated based on the Norwegian

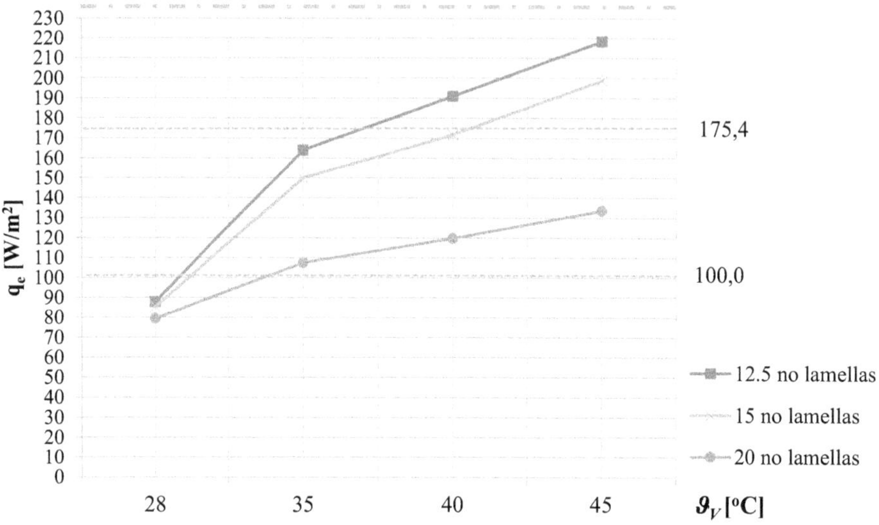

FIGURE 4.8 Dependence of the q_e value on the supply temperature ϑ_V for heating stations **without lamellas** according to standard [8] (Source: Prepared by Karpiesiuk J.), where ----- *q* lower border (in the central zone) and upper border (in the edge zone) — · — *q* limit for a wall surface radiator.

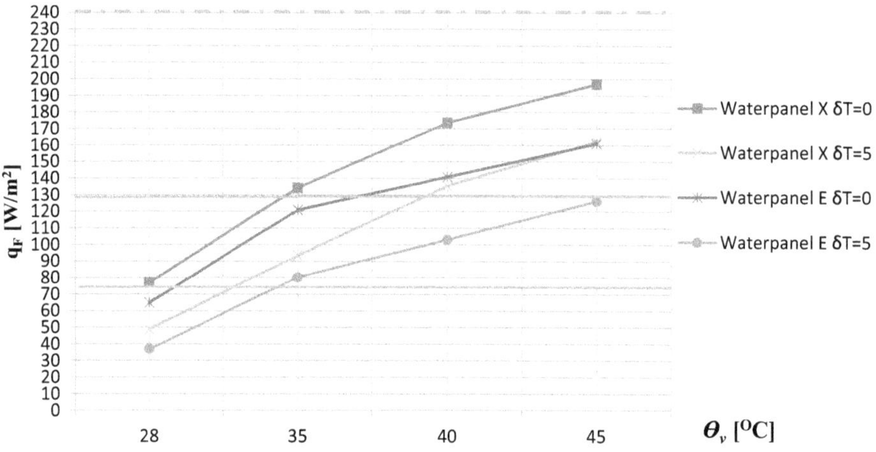

FIGURE 4.9 Dependence of the efficiency q_F on the supply temperature θ_V, with a temperature difference supply and return δT for heating stands *without* lamellas (Waterpanel) (Source: Prepared by Karpiesiuk J.), where ----- *q* lower limit (in the central zone) and upper limit (in the edge zone) for heating coil spacings of 10 and 12.5 cm — · — *q* limit for a wall surface radiator.

standard [1], relative to the coil supply temperature, in the range up to 35°C. At higher supply temperatures, the calculation data is higher than that given in Figure 4.9 from [7]. The comparative values of these heat flux densities are presented in Table 4.2.

TABLE 4.2
The Calculated Value of Thermal Efficiency of the Lightweight Heated Floor *without Lamellas*

	Computational Thermal Efficiency q_{ob} [W/m²]				Thermal Efficiency from the Catalogue q_F [W/m²]			
	Lightweight Underfloor Radiator without Lamellas							
	Heating Coil Supply Temperature [°C]							
Heating Coil Spacing [cm]	28	35	40	45	28	35	40	45
12.5	65.5	122.9	163.8	204.8	64.8	120.9	140.8	160.9
	$q_{12.5}^{max}$ = **73.73**				q_{10}^{max} = **100.00**			

Source: According to the author's guidelines (based on the standard [1]), and the company's catalogue [7].

Graphs of the dependence of thermal efficiency q on the supply temperature, marked as $\vartheta_V = \theta_V$, and regarding underfloor heating without heat-dissipating lamellas in Figures 4.8 and 4.9, are divergent. At the same supply temperature of 28°C, a heating coil spacing of 12.5 cm, efficiency q_e = 87.9 W/m² (Figure 4.8), while in Figure 4.9, q_F = 64.8 W/m². At the 35°C temperature, respectively, q_e = 163.9 W/m² and q_F = 120.9 W/m². The above differences in thermal efficiency q, depending on the heating system supply temperature in the range of 28°C–35°C, amount to approximately 35% and are significant in the calculations. It seems that the thermal efficiency q_e, calculated based on the formula (4.12) from the PN-EN 15377-1:2008 standard [8], is overestimated in the experimental test without lamellas. The reason for these differences can be seen in the fact that the formula (4.12), used for calculations, which applies to heating systems included in the standard [8], does not apply to lightweight underfloor heating without screeds. This formula (from the standard [8]) should be modified to be similar to the formula (4.1) of the standard NT VVS 127:2001 [1], which was developed specifically for lightweight underfloor heating systems. The problem is that it is only concerned with the radiant heaters with heat-dissipating lamellas. Here, the thermal conductivity coefficient of the dissipation layer k_{CL} corrects (reduces) the thermal efficiency flux q. Moreover, based on data from the standard [1], an incorrect entry of one parameter should be eliminated. In the formula for K_H, instead of θ_H^{max} the denominator should be θ_F^{max}. The appropriate form of the formulas from the standard [1] for calculating the theoretical heat flux density should be (4.1) and (4.2) of this monograph.

4.4 TABULAR SUMMARY OF THE THERMAL POWER OF "HEAVY" AND "LIGHTWEIGHT" RADIANT HEATING SYSTEMS

A summary of the results evaluating the computational and experimental thermal power of lightweight floor radiators, based on formulas from European standards, the Norwegian standard, and the original author's calculations, is summarised in Tables 4.3 and 4.4. Additionally, Table 4.5 presents the results of the measurements of the thermal power of "lightweight" systems, with and without lamellas, and "heavy" wet and dry (25 mm thick gypsum screed with a conductivity coefficient of $\lambda \approx 0.3$ W/m·K) with lamellas, of underfloor heating systems for comparative purposes, which were achieved through research and calculations described in the literature, [1–5,9–13]. It refers to the heating coil supply temperatures of 30°C and 35°C, which scientists consider the suggested maximum in LRHS. The data were for a heating coil with a diameter of 16 mm (Purmo pipes have 17 mm [11]), a room temperature of 20°C, and a floor resistance $R \approx 0.00$ m²K/W (for a ceramic tile floor). An approximation of the result was used when direct data were unavailable or when data were obtained using various calculation techniques.

TABLE 4.3
Value of Calculated Thermal Efficiency According to Standard [1] and Experimental Lightweight Heated Floor with *Lamellas*

Coil Spacing [cm]	Computational Thermal Efficiency, q_{ol} [W/m²]				Experimental Thermal Performance, q_{el} [W/m²]			
	Lightweight Floor Radiator with Lamellas							
	Heating Coil Supply Temperature [°C]							
	28	35	40	45	28	35	40	45
12.5	69.2	129.7	172.9	216.2	68.4	136.5	187.3	219.5
15	63.4	118.8	158.4	198.0	62.6	124.1	159.4	177.2
20	51.3	96.3	128.3	160.4	45.1	77.8	88.0	96.1
	$q_{12.5}^{max} = \mathbf{77.81}, q_{15}^{max} = \mathbf{71.28}, q_{20}^{max} = \mathbf{57.75} (\theta_F^{max} = 29°C)$							

TABLE 4.4
Value of the Calculated Thermal Efficiency q_{ob} and Experimental q_{eb} Lightweight Heated Floor *without lamellas*

Coil Spacing [cm]	Calculated Thermal Efficiency, q_{ob} [W/m²]				Experimental Thermal Efficiency, q_{eb} [W/m²]				Experimental Thermal Efficiency, q_e [W/m²]			
	Lightweight Underfloor Radiator without Lamellas											
	Heating Coil Supply Temperature [°C]											
	28	35	40	45	28	35	40	45	28	35	40	45
12.5	65.5	122.9	163.8	204.8	64.8	120.9	140.8	160.9	87.9	163.9	190.9	218.3
15	59.3	111.2	148.2	185.3	57.0	100.0	114.4	132.6	85.4	149.9	171.6	198.7
20	46.9	88.0	117.3	146.6	41.9	56.6	63.2	70.4	79.5	107.4	119.7	133.5
	$q_{12.5}^{max} = \mathbf{73.73}, q_{15}^{max} = \mathbf{66.70}, q_{20}^{max} = \mathbf{52.78} (\theta_F^{max} = 29°C)$								$q^{max} = \mathbf{100.00} (\theta_F^{max} = 29°C)$			

Source: According to author's guidelines based on the standard [1] and q_e according to standard [8].

TABLE 4.5
Comparative Summary of the Thermal Power of Various Types of Underfloor Heating

Coil Spacing [cm]	Thermal Power of Various Types of Underfloor Heating Based on Tests and Calculations [W/m²]													
	"Heavy" Wet Screed with Metallised Foil				"Heavy" Dry Screed with Aluminium Lamellas				"Lightweight" Floor with Lamellas				"Lightweight" without Lamellas	
	[13]		[9,11]		[2,5]		[3]		[3]		[1,4]		[4,10]	
	Heating Coil Supply Temperature [°C]													
	30	35	30	35	30	35	30	35	30	35	30	35	30	35
10	—	90.3	—	97	44	70	36.5	56.5	67.5	104	—	—	60	88
12.5	—	84.1	—	90	42	63	35.0	53.0	63	97	86.5	129	80	121
15	—	77.8	—	84	—	—	33.0	50.0	60	93	79.2	118	69	105
20	—	67.4	—	73	—	—	—	—	—	—	64.1	96	52	72

Source: Based on research and calculations from the literature [2–5,10,11,13] and the standard [1,9] with a heating coil spacing of 10–20cm and a supply temperature of 30°C and 35°C.

Based on the results of Tables 4.3 and 4.4, it seems that the author's calculation methodology for thermal efficiency (with lamellas, defined by formula (4.7) and without lamellas, defined by formulas (4.8) and (4.13)) can meet, or at least come close to, its appropriate results. This is confirmed by the data from both tables, which show that the calculated and experimental efficiencies are similar for heating pipe distances up to 15cm and heating coil supply temperatures up to 35°C. This is the maximum supply temperature for heating pipes to be assumed when installing lightweight heating systems, as confirmed by many studies cited in this monograph. However, one should be careful not to consider the adopted calculation methodology, especially the one based on the European standard PN-EN 15377-1:2008 [8], as fully reliable.

Our original experimental research was conducted in laboratory conditions, without accounting for differences in water temperature between the heating coil outlet and return. The electric heating cable installed inside the heating coil, as permitted by the standard [1] tests, maintained a constant temperature in the pipe. Therefore, it should be borne in mind that the results of laboratory tests may be overestimated compared to natural conditions. This can be observed in Table 4.4 and in the "lightweight" column without lamellas. The results of experimental tests performed with a water-fed heating coil [10], at temperatures of 30°C and 35°C, respectively, had a thermal power of 60 and 88 W/m², with a spacing of 10cm, and the coil powered by an electric heating cable [4], with a larger spacing (12.5cm), instead of having a lower power, had 80 and 121 W/m², respectively. Similarly, the results of numerical calculations [3] of the "lightweight" system with lamellas are lower than those of the experimental results with a heating cable in the heating coil [4]. The thermal power from the "Lightweight" column with lamellas (from Table 4.4) and with a heating coil spacing of 12.5 cm in numerical calculations is 63 and 97 W/m^2 (for 30°C and 35°C, respectively) and in experimental tests [4] much more (86.5 and 129 W/m^2).

Therefore, it is recommended to perform additional experiments, preferably under natural build conditions, on this lightweight surface radiator type, and then make any adjustments to the methodology and mathematical calculations I have adopted to confirm their reliability.

The comparative summary of thermal power from Table 4.5, under the same work conditions, clearly shows that

- Higher thermal power in "lightweight" heating systems is approximately 20%–45% compared to the "heavy" wet screed system with a heating coil spacing of up to 15 cm,
- The "heavy" heating with the wet screed has greater thermal power than "heavy" heating with the dry screed due to the higher thermal conductivity of the screed despite its greater thickness,
- The thermal power of "lightweight" heating with lamellas is at least 50% higher than the power of "heavy" heating with the dry screed.

4.5 TEMPERATURE DISTRIBUTIONS ON THE SURFACE OF LIGHTWEIGHT FLOOR RADIATORS, WITH AND WITHOUT HEAT-DISSIPATING LAMELLAS

Figures 4.10–4.13 show the surface temperature distributions on the surface radiators with and without lamellas, for supply temperatures of 28°C, 35°C, 40°C, and 45°C and different heating coil spacings, based on the author's research described in [4]. The rectangle marked in the middle of the graphs (with a dashed line) indicates the neutral area, regarding the coil spacing of 12.5 cm, due to the different widths of the first heating model, which is 100 cm (50 cm for each part of this first model – with and without lamellas). The stands with heating coil spacings of 15 and 20 cm were 120 cm wide, each with a system width of 60 cm, with or without lamellas.

The average temperature difference in the radiator surface between the heating coil and above it was

- at $\vartheta_V = 28^\circ$ C for heating coil spacings of 12.5, 15, and 20 cm *with lamellas* 0.2°C, 0.3°C and 1.0°C, and *without lamellas* 1.0°C, 0.5°C, and 2.1°C;
- at $\vartheta_V = 35^\circ$ C for heating coil spacings of 12.5, 15, and 20 cm *with lamellas* 0.4°C, 0.9°C, and 6.0°C, and *without lamellas* 2.4°C, 2.5°C, and 7.0°C;
- at $\vartheta_V = 40^\circ$ C for heating coil spacings of 12.5, 15, and 20 cm *with lamellas* 1.0°C, 2.9°C, and 11.7°C, and *without lamellas* 5.2°C, 5.0°C, and 11.5°C;
- at $\vartheta_V = 45^\circ$ C for heating coil spacings of 12.5, 15 and 20 cm *with lamellas* 2.7°C, 5.0°C, and 15.5°C, and *without lamellas* 7.0°C, 8.0°C, and 14.5°C.

For all given heating pipe spacings and when using heat spreaders, we can use this type of heating on walls, as it achieves maximum efficiency without exceeding the maximum permissible temperature of 40°C. The differences between the radiator

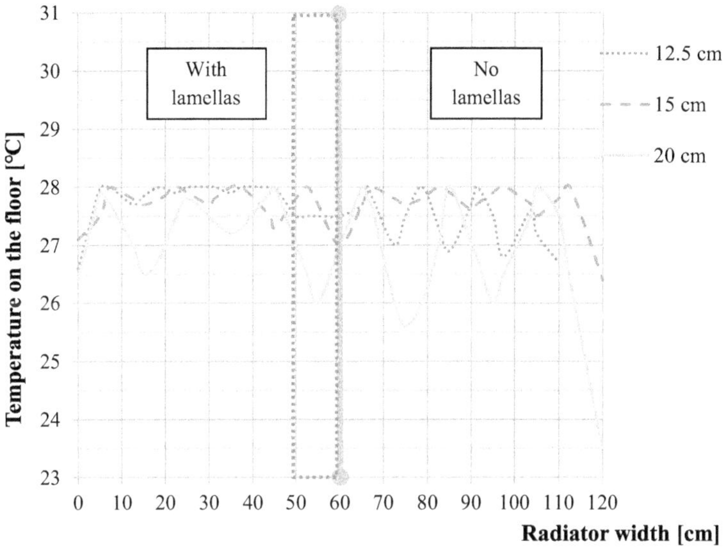

FIGURE 4.10 Temperature profile across the width of the radiator with and without lamellas, at a temperature of 28°C. (Source: Prepared by Karpiesiuk J.)

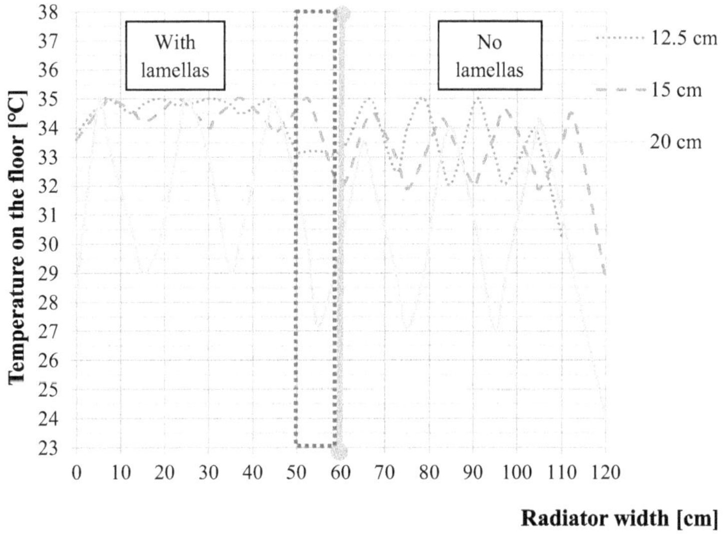

FIGURE 4.11 Temperature profile across the width of the radiator with and without lamellas, at a temperature of 35°C. (Source: Prepared by Karpiesiuk J.)

surface temperature above the heating pipes and between them determine the comfort of use. In this type of unusual radiator, this is important given its thinness. It was assumed that the comfort of use at the maximum permissible floor surface temperature in the central zone, according to the standard [14] of 29°C, is achieved when the average temperature difference between the coil and the floor surface does not exceed

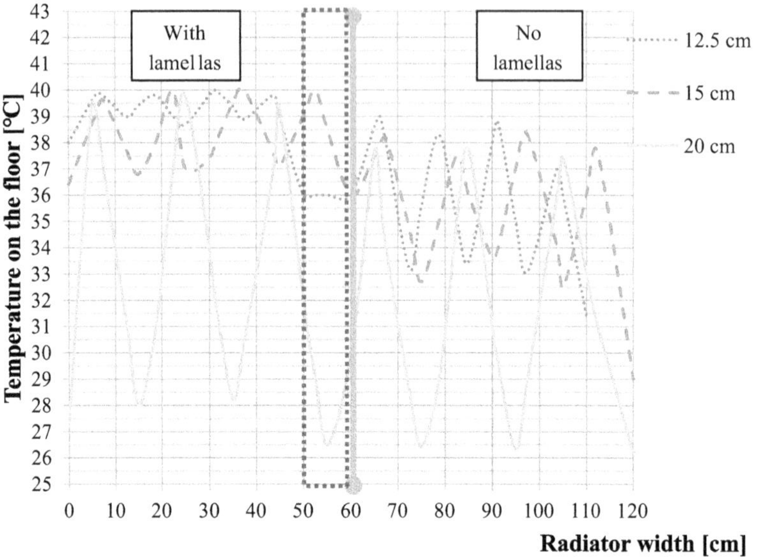

FIGURE 4.12 Temperature profile across the width of the radiator with and without lamellas, at a temperature of 40°C. (Source: Prepared by Karpiesiuk J.)

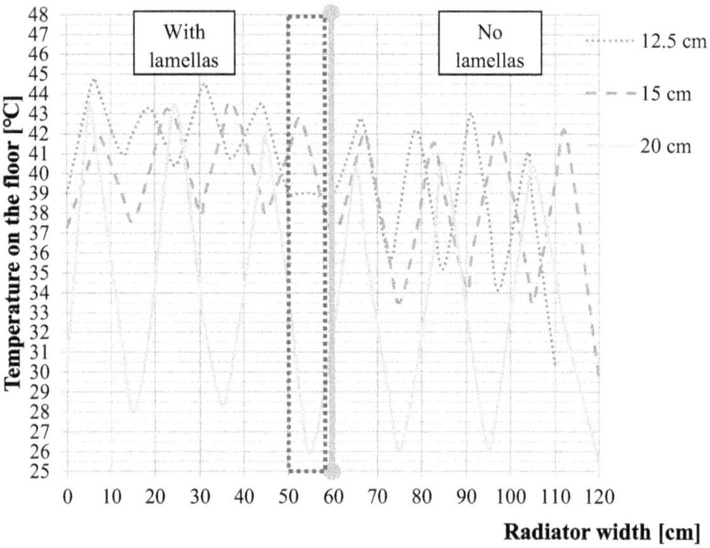

FIGURE 4.13 Temperature profile across the width of the radiator with and without lamellas, at a temperature of 45°C. (Source: Prepared by Karpiesiuk J.)

2 K. Accordingly, in the edge zone at a maximum temperature of 35°C, this difference will not exceed 2.5 K. With this in mind, a supply temperature of 28°C–35°C and a pipe spacing of 12.5–15 cm can be considered comfortable for a lightweight radiator with lamellas, provided the average difference between the temperatures

above the heating coil and between them is 0.2°C–0.9°C. A radiator without lamellas will be comfortable with an average temperature difference of 1.0–2.5 K under the above assumptions, with a heating coil spacing of up to 12.5 cm. The 20 cm spacing between the pipes is comfortable only when using heat spreaders (average temperature differences will be 1.0–3.8 K). So, the 20 cm spacing of the heating coils eliminates the possibility of this type of radiator without lamellas, as the resulting temperature differences on the flooring negatively affect human physiology.

Similar results regarding the temperature distribution on the floor surface, reported by Dongliang et al. [15] and Karpiesiuk and Chyży [4], are also reported by Karpiesiuk [16]. The average temperature on the ceramic tile floor surface above and between the heating coils, at a spacing of 10–15 cm, is given in Tables 4.6 and 4.7, according to [16].

A suggested selection of an ultra-thin surface radiator, taking user comfort into account, is shown in Table 4.8. It is based on experimental studies and the assumption that at the maximum permissible floor surface temperature of 29°C in the central zone, as per PN-EN 15377-1:2008 [8], the average temperature difference above and between the coils will not exceed 2 K. Similarly, in the peripheral zone, at a maximum temperature of 35°C, this difference will not exceed 2.5 K.

4.6 SUMMARY OF LABORATORY RESEARCH, THERMAL POWER CALCULATIONS, AND COMFORT OF LRHS USE

As the experimental results showed, a lightweight radiator has low inertia. It seems that the calculation methodology for thermal efficiency with lamellas and the original author's calculation methodology without lamellas, respectively, defined by formulas (4.1) and (4.8), can meet or at least come close to their appropriate results. This

TABLE 4.6
Temperature Differences on the Flooring Ceramic Tiles for the Radiant Heaters with Lamellas

	With Aluminium Lamellas			
Model	**Temperature, θ_V [°C]**	**Room Temperature [°C]**	**Pipe Spacing [cm]**	**Average Temperature Difference in the Radiator Surface Above the Heating Coil and between It, $\delta\theta_V$ [°C]**
XPS	28	20	15	0.3
Insulating	35			0.9
Boards	40			2.9
	45			5.0
EPS Insulating	28	20	12.5	0.2
Boards	35			0.4
	40			1.0
	45			2.7

Source: According to [16].

TABLE 4.7
Temperature Differences on the Flooring Ceramic Tiles for the Radiant Heaters without Lamellas

Without Aluminium Lamellas				
Model	**Temperature, θ_V [°C]**	**Room Temperature [°C]**	**Pipe Spacing [cm]**	**Average Temperature Difference in the Radiator Surface above the Heating Coil and between It, $\delta\theta_V$ [°C]**
XPS	28	20	10	0.9
Insulating	35			2.3
Boards	40			3.6
	45			4.5
EPS Insulating	28	20	12.5	1.0
Boards	35			2.4
	40			5.2
	45			7.0

Source: According to [16].

TABLE 4.8
Selecting an Ultra-Thin Surface Radiator to Achieve Comfort, Depending on the Pipe Spacing

Heating Coil Spacing	The Centre Zone of the Floor	The Edge Zone of the Floor	The Wall
12.5 cm with lamella	+	+	+
15.0 cm with lamella	+	+	+
20.0 cm with lamella	+	–	+
10.0 cm without lamella	+	+	+
12.5 cm without lamella	+	+	+
15.0 cm without lamella	+	–	+
20.0 cm without lamella	–	–	+

+ recommended; – not recommended.

is confirmed by the data in Tables 4.3 and 4.4, where the computational and experimental efficiencies, based on their own calculation assumptions, are similar over heating pipe distances up to 15 cm and supply temperatures up to 35°C. However, one should be cautious about considering the adopted original calculation methodology, and even more so the one based on the European standard PN-EN 15377-1:2008 [8], as fully reliable. It is recommended to conduct additional experiments on this type of lightweight surface radiator and then make any necessary adjustments to the methodology and mathematical calculations I have adopted to confirm their reliability.

When we compare Figure 4.6 for underfloor heating with lamellas, developed based on experimental research and using the formula of the Norwegian standard [1] for lightweight underfloor heating systems, with Figure 4.7, produced by the manufacturer from [7], the thermal power results are similar (Table 4.1).

However, the comparison of the thermal power values determined in the experiment, based on the European standard [8], which does not take into consideration the correction factor k_{CB}, and the author's calculations, where this coefficient is taken into account, clearly shows that the use of the k_{CB} factor contributes to the credibility of the heat flux density value q without lamellas.

Thus, at temperatures of 28°C and 35°C, with a spacing of 12.5 cm, the efficiency q is as follows:

- when includes $k_{CL} \rightarrow q_{eb} = 64.8$ and 120.9 W/m^2 (Table 4.4);
- when does not include $k_{CL} \rightarrow q_e = 87.9$ and 163.9 W/m^2 (Table 4.4).

The thermal power q_{eb} is then close to the theoretical calculations with a heating coil spacing of 12.5 cm, where $q_{eb} = 65.5$ W/m^2 (28°C) and 122.9 W/m^2 (35°C). The research results confirmed that in the middle zone, both a lightweight underfloor heater with lamellas, where the heating coil spacing is up to 20 cm, and one without lamellas, where it is up to 15 cm, provide a comfortable temperature distribution on the ceramic floor. For this reason, it would be worth developing a standard formula covering not only the use of lightweight heated floors with heat spreaders, as per the Norwegian standard [1], but also without them, as it should not rely on European standard models. European standards do not provide formulas for calculating this specific type of floor without screeds or heat-dissipating lamellas. Until a standard methodology for LRHS calculations without lamellas is developed, it is recommended that the author's calculations using formula (4.8) with a spacing of heating pipes up to 15 cm be used.

Thermal efficiency values calculated according to the standard [1] (necessary after applying corrections to the formulas presented by the author of this monograph) can be considered reliable. Both the values from the standard formulas for theoretical calculations and the calculation results based on experimental tests do not differ significantly for heating coil spacing up to 15 cm (excerpt from Table 4.3):

12.5 cm	69.2	129.7	172.9	216.2	68.4	136.5	187.3	219.5
15 cm	63.4	118.8	158.4	198.0	62.6	124.1	159.4	177.2

Greater thermal efficiency differences occur with a heating coil spacing of 15 cm and a supply temperature of 45°C, which is not that important, because already at a supply temperature of 28°C–29°C a lightweight radiator with lamellas reaches the maximum permissible temperature on the floor. Only when the heating coil spacing is higher (20 cm) are the thermal efficiency differences significant, and those calculated using standard formulas are overestimated compared to those calculated on the basis of tests (excerpt from Table 4.3):

20 cm	51.3	96.3	128.3	160.4	45.1	77.8	88.0	96.1

Caution is recommended when designing lightweight underfloor heating based on the Norwegian standard [1] when the heating coil spacing is greater than 15 cm. Experimental tests indicate that the heat flux density, with a heating pipe spacing of 20 cm, should be lower than standard calculations by about 10%–20% when the supply temperature is not higher than 35°C.

Based on the thermal output results for "lightweight" underfloor heating systems, as presented in Table 4.4, numerical and experimental studies, it is recommended to conduct additional experiments and calculations to address the significant discrepancies in the results. Experimental testing of the lightweight radiant heating/cooling system (LRH/CS) should be performed outside the laboratory under "in situ" conditions with a water-fed coil. The entire system should be connected to a reversible heat pump, and heating and cooling measurements should be taken. The experimental results should then be calculated numerically (e.g., using ANSYS) and additionally in accordance with the Norwegian standard [1]. Such an experiment, supported by numerical and mathematical calculations, will allow verification of the thermal output and comfort values of the LRH/CS and confirm the calculation methodologies of existing standards and the author's own methodology for the "lightweight" system without lamellas.

Summarising the results of experimental tests, calculations of thermal parameters, analysis of standard data, and based on other literature sources mentioned in this monograph, regarding the LRH/CS, also called Ultra-Thin Radiant Heating/Cooling System, it can be concluded that:

- This floor construction, both with and without metal heat spreaders, can be an effective emitter of heat or cold. One only needs to use appropriate distances between the coils on the floor in the middle zone, where people stay, preferably not exceeding 20 cm for radiators with elements that improve thermal conductivity and up to 15 cm for radiators without these elements. Only then will we achieve adequate heating or cooling efficiency and comfort of use, as evidenced by an even temperature distribution on the floor surface.
- In addition to the methodology for calculating the thermal efficiency of a lightweight underfloor radiator with lamellas, an original methodology has been developed for calculating this type of radiator without lamellas, which can be used with the heating coil with a maximum distance of 15 cm (until standards for this lightweight system, which do not require lamellas, are created). The PN-EN 15377-1:2008 [8] standard does not apply to lightweight underfloor heating without screeds and should not be used in this application due to significant discrepancies in results, as shown in Table 4.4.
- Research on the U_{total} heat transfer coefficient shows that the most universal surface, lightweight heating, and cooling system is the one mounted on the wall. Only if the installation site is a wall will high thermal efficiency be ensured in the winter and cooling efficiency in the summer, and additional outlays be avoided. It should be remembered that the location of the radiant heating and cooling system is not the only condition for achieving the highest

efficiency. When designing, it is worth applying the other criteria listed in Section 3.2.

- The results of experimental tests show that the ultra-thin radiant heating system provides the maximum designed temperature on the floor of 29°C at a low supply temperature of up to 35°C, even without heat dissipation lamellas. However, by using dissipative lamellas, we can achieve the standard maximum floor temperature of 29°C, even at a supply temperature of about 30°C. These parameters validate this lightweight radiant system for its effective use in primary energy recovery devices such as heat pumps, photovoltaic cells, and others.
- Additional advantages of LRH/CS:
  - Its low weight allows it to be installed on ceilings with low load-bearing capacity (e.g. wooden) or where self-weight loads need to be reduced;
  - It can be installed directly on wood-based floors (e.g., OSB boards);
  - Its small thickness is important when renovating or modernising rooms, due to the established dimensions of window openings and door lintels;
  - The design allows for quick, easy, and aesthetic installation of heating coils in ready-made grooves of insulating boards;
  - It ensures the cleanliness of assembly and eliminates the use of heavy construction equipment (concrete mixers, units for mixing concrete mortar, etc.);
  - It eliminates the need to wait for the concrete setting process, which speeds up the commissioning of the investment;
  - The low inertia of the system allows for the quick achievement of the desired temperature on the floor surface (a temperature change of 10°C is achieved within a maximum of 1 hour in a radiator design without lamellas, with a heating coil spacing of 10 cm);
  - It is ideal for underfloor heating devices using renewable energy sources (RES), such as heat pumps or other energy carriers with a low heating medium temperature of up to 35°C;
  - For small areas, it may be cheaper than a traditional heating system with different types of screeds, taking into account the material used and labour costs.
- The versatility of using LRH/CS for efficient heating and cooling, along with many other advantages of an ultra-thin surface radiator without screeds, confirmed by research, may make this type of structure popular and widely installed in the coming years. The description and presented test results show that the lightweight radiant heating and cooling system is constantly being developed, improved, and modified to achieve higher thermal comfort and efficiency of this system.

According to the construction law [17], one of the main requirements for designing and building, as per technical and construction regulations and in accordance with the principles of technical knowledge for various types of facilities, is to ensure the safety of use. The condition of safe use is related to the mechanical strength of this type of floor. This means that, in addition to thermal and comfort conditions, this type of heating and cooling system should be tested for mechanical strength.

Such studies were conducted and described in [18]. This confirms that the basic requirements of construction law regarding operational safety, which come next to structural safety, fire safety, health, and environmental protection, are ensured.

ABBREVIATIONS

EPS	expanded polystyrene
LRH/CS	lightweight radiant heating and cooling systems
LRHS	lightweight radiant heating system
RES	renewable energy sources
XPS	extruded polystyrene

REFERENCES

1. NORDTEST Method NT VVS127: UDC 697.1: Approved 2001-11. Floor heating systems. Design and type testing of waterborne heat systems for lightweight structures. Finland.
2. Karpiesiuk P., *Research on the temperature field during surface heating.* Engineering Diploma Thesis, Faculty of Construction and Environmental Engineering, Department District Heating, Heating and Ventilation, 2015. (in Polish).
3. Werner-Juszczuk A.J., The influence of the thickness of an aluminum radiant sheet on the performance of the lightweight floor heating. *Journal of Building Engineering*, 44 (2021), 102896.
4. Karpiesiuk J., Chyży T., *Results of experimental tests on the temperature distribution of surface radiators of lightweight construction without screeds.* Rynek Instalacyjny, November 2017, No. 11/2017, year XXV, pp. 45–50. (in Polish).
5. Werner-Juszczuk A.J., The calculations of the efficiency of the lightweight floor heating according to Nordtest method NT VVS 127. *Innovations – sustainability – modernity – openness. Modernity in engineering.* Edited by Dorota Anna Krawczyk, Iwona Skoczko, Ewa Szatyłowicz, Series of monographs, Volume 43. Printing House of Bialystok University of Technology, Bialystok, 2021, pp. 159–172. ISBN 978-83-66391-90-1 (eBook). https://doi.org/10.24427/978-83-66391-90-1
6. Weitzmann P., Svendsen S., Method for calculating thermal properties of lightweight floor heating panels based on an experimental setup. *International Journal of Low Energy and Sustainable Buildings*, 3 (2005), pp. 1–15.
7. Elektra Kardo technical catalogue, Poland, January 2023.
8. EN 15377-1:2008, ITD. Heating systems in buildings: design of embedded water-based surface heating and cooling systems. Part 1: determination of the design heating and cooling capacity. European Committee for Standardization (CEN): Brussels, Belgium.
9. EN 1264-2: 2008+A1:2012, IDT. Water-based surface-embedded heating and cooling systems. Part 2: floor heating: prove methods for the determination of the thermal output using calculation and test methods. European Committee for Standardization (CEN): Brussels, Belgium.
10. Żukowski M., Karpiesiuk P., Test results of a radiant radiator with a very low height. *Instal*, 10 (2015), pp. 38–41. (in Polish).
11. Technical guide. PURMO POLAND surface heating and cooling systems, 03/2020. Poland. (in Polish).
12. Karpiesiuk J., A new dimension of radiant heating/cooling without screeds (1). Radiator without screeds. *Magazyn Instalatora*, January 2018, No. 1 (233), pp. 34–35. (in Polish).

13. Radiant heating and cooling boards of the KAN-therm system. 03/2022. Poland. (in Polish).
14. EN 15377-2:2008. Heating systems in buildings: design of embedded water based surface heating and cooling systems. Part 2: design, dimensioning and installation. European Committee for Standardization (CEN): Brussels, Belgium.
15. Dongliang Z., Ning C., Zijie W., Experimental and numerical analysis of lightweight radiant floor heating system. *Energy and Buildings*, 61 (2013), pp. 260–266.
16. Karpiesiuk J., A new dimension of screed-free heating and cooling (2). *Magazyn Instalatora*, 2 (234, February 2018, pp. 24–25. (in Polish).
17. Act of July 7, 1994 Construction Law. Prepared on the basis of: i.e. Journal of Laws of 2017 pos. 1332, 1529, of 2018, item 12, 317, 352, 650. (in Polish).
18. Karpiesiuk J., *Computable model of the lightweight layered floor loaded by temperature.* Doctor Dissertation, Technical University of Bialystok, 2021. (in Polish).

5 Thermal Power of Various Flooring Types in a Lightweight Radiant Heating System

A lightweight radiant heating system (LRHS) can be used with various types of flooring. Most require a perfectly level substrate. This can be achieved by using thin, dry screeds or self-levelling screeds, reinforced with a composite of adhesive-mesh layer. These subfloors constitute the load-bearing layer of the LRHS. This chapter calculates the thermal output (heat flux density) of eight flooring variants with three thin, dry screeds and one variant with a self-levelling screed. Calculations of the thermal output of lightweight floors accounted for the use of heat dissipators (lamellas) made of aluminium foil. Expanded polystyrene (EPS) or extruded polystyrene (XPS) boards provided thermal insulation under all the screeds. The thermal output results are presented in the tables. They indicate that heat-dissipating lamellas, spacing, and coil temperature significantly increase the thermal output (efficiency) of the heated floor. Flooring materials or the supporting layers of the lightweight floor structure have a smaller impact on thermal efficiency. The thermal output results presented in this chapter for the most commonly installed floors can be used to design lightweight heated floor systems with or without thin screeds. The thermal output values in the tables, derived from mathematical calculations, are reliable. However, you should confirm through experimental testing.

The thermal output of lightweight heated floors is the most reliable determinant of room comfort in various types of buildings. Depending on the room type, its function, user preferences, or aesthetic value, different types of flooring are used on heated floors. This chapter selects both commonly installed and less frequently installed flooring. Calculations of the thermal output of ceramic tile floors, whose thermal efficiency was calculated in Chapter 4, are omitted. The choice of different types of flooring or coverings is an important issue, given the increasing number of surface heating systems installed in various aesthetic and structural configurations of floors, walls, and ceilings, not only in the affluent continent of Europe and North America, but also in Asia and the Pacific, where the market for heating and cooling systems is growing. Among other places, China is experiencing one of the world's highest growth rates in surface heating installations. This is particularly influenced in European countries such as Germany, the United Kingdom, and Scandinavian countries by the development of modern, energy-efficient, passive, zero-energy, or even energy-plus buildings, driven by rising energy costs and environmental protection. In North America, investments in radiant heating systems have exceeded

DOI: 10.1201/9781003679004-5

$1 billion and continue to grow. This is particularly important in the northern, colder regions of the United States, where climate conditions vary throughout the year, as confirmed in [1].

This chapter presents mathematical calculations and a comparative analysis of the thermal properties of various floor types used in lightweight underfloor heating structures, based on the engineering thesis [2], supervised by Jacek Karpiesiuk. The materials for lightweight underfloor heating were selected to meet the technical requirements for their use, i.e., to maintain load-bearing capacity under loads in multi-family buildings on rigid insulating substrates with a strength of 200 kPa (EPS) and 300 kPa (XPS). Various coil spacings, ranging from 10 to 20 cm, were also used, which is an important factor influencing thermal efficiency. It is confirmed by mathematical models, numerical calculations, and experimental studies [3]. These also indicate that changing the coil spacing does not significantly affect the uniformity of heat dissipation when heavy screeds (6–8 cm) are used. The coil, as a heat emitter, located at the bottom of the screed, can distribute the heat evenly, thanks to the concrete layer at least 4.5 cm thick above it. However, this is important when using thin screeds, or no screed at all, as they hinder even heat distribution on the floor. The so-called cold spots may appear between the heating pipes, especially when their spacing is significant – over 20 cm – as confirmed by the user comfort study and described in Chapter 4. The thermal outputs (heat flux density) for many variants of light underfloor heating, with or without dry screeds, were calculated and presented in the tables in this chapter. A crucial aspect of thermal efficiency calculations is the consideration of various specific flooring types in the LRHS. Most analyses assume that the thermal efficiency depends mainly on the flooring thermal resistance values, providing only their ranges [4]. Some catalogues entirely skip the impact of the finishing layer on the actual thermal efficiency, even though this is known to be an essential parameter. Floor thickness, the material it is made of, and the method of installation significantly affect the amount of heat transferred to the rooms. This chapter conducts a multidimensional analysis of a lightweight heated floor, considering eight flooring types, three thin dry screed types, a floor system without a screed, with or without lamellas, and a maximum 20 cm spacing between heating pipes to ensure user comfort. Additionally, the principle of minimising investment costs while maintaining the lightweight flooring installation technology was followed.

5.1 OUTPUT DATA WITH THERMAL POWER CALCULATION PROCEDURE

To calculate the thermal power of various types of LRHS floors, the procedures and calculation formulas described in Chapter 4 were used, based on the Norwegian standard [5] for a floor system with aluminium heat-dissipating lamellas. One of the formulas in the Norwegian standard was improved by Karpiesiuk [6], who additionally developed a procedure and formulas for calculating the thermal power of LRHS without metal lamellas, based on the aforementioned Norwegian industry standard. Modifications to the standard [5] were introduced due to significant discrepancies, even over 50%, in heat flux density values compared to European standards and numerical calculations, as confirmed by the author of this monograph in his article [6] and in another article [7]. These comparative analyses of the NT VVS 127 method versus other calculation methods for determining thermal power parameters for lightweight underfloor heating systems revealed significant discrepancies between the results obtained using the Nordtest method and the tests or numerical calculations. It is also essential to amend the Norwegian standard and add a new thermal power calculation procedure that accounts for the absence of metal lamellas in the LRHS. It therefore seems reasonable to apply the corrections to the formula in the Norwegian standard, as well as the procedure for calculating the thermal power of a lightweight floor system without lamellas, developed based on the work by Karpiesiuk [6]. The author of this monograph has been in correspondence with a representative of the Scandinavian scientific community who has been legalising the Nordest standards for 2 years regarding this correction. Unfortunately, to date, there has been no binding decision to correct the error contained in the standard. Furthermore, it is worthwhile to conduct experimental studies to assess the validity of the heat flux density calculation procedure proposed by Karpiesiuk [6] in LRHS with and without heat-dissipating lamellas.

5.1.1 Materials and Flooring Cover Used in the LRHS

When designing floors with different types of flooring, several criteria described earlier were considered. Still, the most important thing was ensuring that floors requiring a level surface had a perfectly constructed load-bearing subfloor. Furthermore, the selection of floor structures was guided by the principle of limiting screed height while meeting the manufacturers' load-bearing requirements. The design process for floors with different types of flooring aimed to combine the advantages of lightness and compare the calculated thermal outputs, while also applying all installation procedures suggested by the manufacturers. As is well known, the Nordest standard should not be used to calculate the thermal performance of floors with both dry and wet screeds. Therefore, floor layers were generally chosen with thickness kept to a minimum while maintaining the installation technology. In one of many design variants, a lightweight floor system was implemented without a screed, but with a thin self-levelling screed as a levelling base for the proposed flooring. This option reduces investment costs for investors implementing the project themselves (without hiring specialists). However, in this option, due to the need to connect the EPS or XPS thermal insulation layer with the supporting adhesive-mesh layer, metal lamellas could not be installed for technological reasons. To reduce investment costs, inexpensive but proven C2S1 cement adhesives were used, reinforced with fibreglass mesh and covered with a self-levelling screed. These components formed the floor's load-bearing layer.

In summary, a multi-variant system of lightweight heated floors was ultimately implemented. The top layer of the floor consisted of eight different types of flooring, designated by manufacturers for use on floors subject to thermal stress. Three different types of thin, dry screeds were also used for analysis. These differ in thickness and thermal conductivity. One of the floor design variants did not have a screed layer as the floor's supporting layer. This was replaced with a proprietary solution comprising a layer of flexible C2S1 cement adhesive reinforced with a fibreglass mesh, and a self-levelling screed covering this composite to create a perfectly even substrate for the eight flooring cover types. The adhesive-mesh layer was directly bonded to the thermal insulation, which consisted of EPS or XPS insulation boards. Dry screeds were laid in one variant directly on EPS or XPS thermal insulation, or, in another variant, separated by a layer of aluminium lamellas to improve heat distribution and increase the thermal efficiency of lightweight heated floors. Heating pipes were placed in the thermal insulation grooves at spacings of 10, 12.5, 15, and 20 cm. These coil spacings were assumed to be the most effective based on previous research by many authors, including [8–12], which tested and analysed LRHS.

Eight different types of flooring cover were selected, along with adhesive layers designed explicitly by the flooring manufacturers for use in thermally stressed flooring systems. The flooring components and their technical data, including the thermal conductivity coefficient λ and thickness, necessary for calculating thermal power, are listed in Table 5.1.

The most commonly used flooring for underfloor heating is the one that conducts heat best. These include various types of stone tiles, such as ceramic tiles, stoneware, natural stones (marble and granite), engineered stones, or mineral-resin composites. Artificial stone composites can be conglomerates, engineered stones, or mineral-resin composites. They are manufactured from natural aggregates (e.g., quartz and marble), bonded with polyester or acrylic resin, and pigmented. The thermal conductivity of all natural or engineered stone floors is high, exceeding 1.0 W/(m·K) according to calculations and standard tables [25,26]. Since the thermal performance of ceramic tiles was calculated in Chapter 4, this chapter only analyses non-stone floors.

Non-stone flooring materials that have been used for many years were selected, such as laminate flooring, PVC flooring, carpet flooring, and glued multilayer

TABLE 5.1
Material Parameters of Eight Flooring Cover Composites [2]

Lp.	Adhesive-Backed Flooring	λ [W/(m×K)]	Thickness [mm]
1.	PVC floor panels – vinyl [13] + polyurethane adhesive	0.45	5.0
	PU 560 [14]	0.25	2.0
2.	PC floor panels – vinyl with integrated XPO underlay [15]	0.21	6.0
3.	Three-layer wooden board [16] + polyurethane adhesive	0.14	10.0
	PU 560 [14]	0.25	2.0
4.	Arendal laminate flooring [17] + Smart Underlay [18]	0.15	9.0
		0.034	3.0
5.	Microflex two-component microcement [19]	0.55	2.0
6.	Eternal PVC flooring [20] + Eurocol 640 Eurostar Special	0.25	2.0
	adhesive [21]	0.25	2.0
7.	Flotex Colour carpet [22] + Eurocol 640 Eurosafe Special	0.06	4.3
	adhesive [21]	0.25	2.0
8.	Sikafloor elastic resin floor – 300 N [23] + Sikafloor 156	0.28	2.0
	primer [24]	0.25	2.0

wooden boards. Vinyl flooring, which has recently been most frequently installed over underfloor heating systems, was also used. Vinyl flooring was chosen for its aesthetic and functional qualities, as well as ease of installation. All these flooring materials must be resistant to thermal stress and must not release harmful volatile compounds at elevated temperatures to be safe for use in occupied buildings. This is important because chemical additives are present in all non-stone flooring materials, including carpet, PVC, vinyl, and wood composites bonded with adhesives. Example symbols for approved floor types for use with underfloor heating systems are shown in Figure 5.1 [27].

When selecting flooring exposed to thermal action, it is important to consider not only its appearance and ease of use, but above all, its thermal properties, which significantly affect heating and cooling efficiency.

The following thin, dry screeds were used to calculate the thermal power:

- BAUNORM Fibercement Board is a 10-mm-thick, high-strength cement-fibre board. This is achieved by combining a specially formulated cement mixture with polypropylene fibres. When properly installed, including glueing to the substrate and between the board edges, this thin composite is recommended by the manufacturer for use on heated floors with a load capacity acceptable in residential buildings.
- Knauf Brio 18 mm gypsum-fibre board. The manufacturer designates this thin gypsum board for use in underfloor heating systems, thanks to the special gypsum mixture reinforced with glass fibres, and the special overlapping of the board edges, which are additionally secured with screws or dedicated staples during installation. For installation details, refer to the manufacturer's technical data sheet (Table 5.2). The combination of this thin dry screed's high quality and a developed installation technique ensures adequate load-bearing capacity, stiffness, and stability, even on substrates

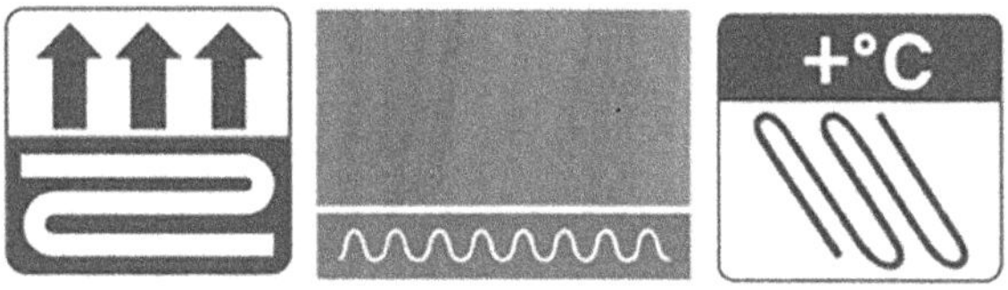

FIGURE 5.1 Example symbols placed on flooring materials for underfloor heating [27].

TABLE 5.2
Material Parameters of Three Dry Screeds [2]

Lp.	Dry Screed	Thermal Conductivity Coefficient, λ [W/(m×K)]	Thickness [mm]
1.	BAUNORM Fibercement Board [28]	0.2166	10
2.	Knauf Brio [29]	0.2500	18
3.	Knauf Dual Floor [30]	0.2500	25

less stiff than concrete. A disadvantage of Knauf Brio gypsum-fibre board is its significantly higher price compared to BAUNORM Fibercement Board.

- Knauf Dual Floor plasterboard, 25 mm thick. This is the thickest dry screed included in the analysis. Installation is relatively simple, involving glueing or stapling two 12.5-mm-thick plasterboards. The plasterboards are internally reinforced with glass fibre, which increases their bending resistance and improves the stability of the entire system. Despite being thicker than other dry screed boards, Knauf Dual Floor plasterboard is the most affordable solution for installation.

Table 5.2 lists the most important parameters of the three dry screeds mentioned, which are necessary to calculate the thermal power LRHS.

A multivariate analysis of the thermal output of lightweight heated floors was based on eight selected floorings and three screeds with heat-dissipating aluminium lamellas. A floor layout was also adopted that omitted the use of any screed and did not install aluminium lamellas, due to the possibility of easily combining the thermal insulation layer with the deformable cement adhesive layer. Generally, three lightweight heated floor layer layouts were used to calculate the thermal output.

The first heated lightweight floor layout with metal lamellas consisted of the following layers:

- eight flooring options with dedicated adhesive
- three screed options
- a multilayer heating pipe, 16 mm thick with an aluminium insert
- a 90-μm-thick aluminium heat-dissipating lamella
- two options of EPS and XPS thermal insulation with the same thickness of 3 cm

The second heated lightweight floor layout without metal lamellas consisted of the following layers:

- eight flooring options with dedicated adhesive
- three screed options
- a multilayer heating pipe, 16 mm thick with an aluminium insert
- two options of EPS and XPS thermal insulation with the same thickness of 3 cm

The third heated lightweight floor layout without screeds and metal lamellas consisted of the following layers:

- eight flooring options with dedicated adhesive
- a 5-mm-thick self-levelling cement layer
- a composite of deformable cement adhesive and 3-mm-thick fibreglass mesh
- a multilayer heating pipe, 16 mm thick with an aluminium insert
- two options of EPS and XPS thermal insulation with the same 3 cm thickness

When calculating the floor's thermal output, heat losses under thermal insulation were omitted, assuming that the air temperature in the heated room and beneath it was identical. Thermal insulation layers with milled grooves were adopted from the manufacturer's catalogue and are boards made of XPS – Waterpanel X and EPS – Waterpanel E [31]. Their primary function is to limit heat loss to the lower floor. Aluminium heat-dissipating lamellas [32] are arranged in milled grooves in the EPS and XPS thermal insulation. Throughout the entire floor system, they have the highest thermal conductivity coefficient but the lowest layer thickness. The high thermal conductivity of metal lamellas can be a significant factor in improving heat flux density and user comfort by enhancing heat dissipation. This is particularly important in lightweight heated and cooled floor systems characterised by low weight and low thermal inertia. The water in the heating coil, with an aluminium insert [33], is connected to the heating circuit and serves as the heat source.

The advantage of the third layout without screeds is the thinness of the carrier layer, consisting of an adhesive-mesh composite and a self-levelling screed, at only 8 mm. Despite the absence of metal lamellas in this system, the good thermal conductivity of the carrier layer can enable a relatively high heat flux density of the heated lightweight floor. Furthermore, this is the only one of the three analysed layouts that does not contain screeds in accordance with the requirements of the Scandinavian standard [5].

5.1.2 Determination of the Design Thermal Power of an Underfloor Heating Layout with Thin Screeds and Lamellas

The first heated lightweight floor layout with thin screeds and metal lamellas was subjected to mathematical calculations using formulas adopted from the standard [5]. This subsection presents an example calculation of the heat flux density (thermal power) for a specific configuration of selected layers from the first layout. The example calculations assumed a floor made of glued vinyl panels. This floor covering is bonded to a 10-mm-thin screed made of BAUNORM Fibercement Board fibre-cement boards using a polyurethane adhesive designed by the vinyl panel manufacturer. This installation method significantly improves heat transfer. The absence of air gaps to prevent the free flow of heat is the recommended layout. The screed was laid loosely (without being glued) on a thermal insulation layer covered with aluminium lamellas. Heating coils were placed in the grooves in the thermal insulation, acting as a heat source. Because the dry screed slab is not glued to the thermal insulation, it must be assumed that the resistance of the conductive layers is greater than that of a carefully shaped, permanently glued layer. This is because, without glueing, an air gap appears, acting as a thermal barrier. It is possible to glue the screed to the aluminium foil substrate. Still, this procedure is expensive and rarely used, as installation requires either expensive adhesives, such as polyurethane or epoxy, or a labour-intensive passivation, activation, and impregnation technique, which allows for the use of less expensive, deformable cementitious adhesives. This layer configuration from the first layout was chosen due to the current high popularity of vinyl flooring in underfloor heating systems and the use of the thinnest 10 mm screed.

This layer arrangement is consistent with the recommendations of the standard [5] for calculating heat flux density in a lightweight heated floor system. Whereas, a loose arrangement of layers between the flooring cover and screed is found, for example, in floor panels connected using tongue-and-groove or click joints without glueing to the substrate. This arrangement was also adopted for calculations and is listed under item 2 in Table 5.1, where PC vinyl floor panels with an integrated XPO underlay are laid loosely on a dry screed, and in item 4 of Table 5.1. Here, a different type of panel was used – laminated Arendal panels with a Smart Underlay, laid without glueing on a dry screed. This type of flooring would be best laid directly on metal lamellas to improve heating efficiency. The calculation example allows similar calculations of the thermal output for other variants of the first-floor layout with different flooring, screeds, thermal insulation, and heating pipe spacing.

Similar to the heat flux density calculations in Chapter 4 for ceramic floors without screeds, the same calculation model was adopted to simplify the thermal analysis. In this model, the heat transfer process occurring within the lightweight floor structure was reduced to a simplified system of three equivalent thermal resistances. These are R_{HC} – the thermal resistance from the heating medium to the heat-conducting layer – the metal lamellas; R_i – the thermal resistance of the heat-conducting layer to the heated room towards the floor; and R_e – the thermal resistance beneath the heat-conducting layer towards the room below or outside. These resistances define the individual floor layers, and additionally, the heat transfer resistances on the internal and external surfaces were taken into account. This type of calculation model was adopted due to the high computational complexity of heat flow. The internal room (air) temperature was assumed to be 20°C. This mathematical model is sufficient to precisely determine the heat flux density transferred from the floor to the heated room. In addition to the thermal power, the maximum water temperature in the coil was also determined in the described example variant based on the formulas from the standard [5]. All components of the formulas given below were already listed and described in Chapter 4, so a detailed description of their meanings is not required. Furthermore, a description of the symbols is provided at the beginning of the monograph. Examples of heat flux density and water temperature calculations in the coil for three heated floor layouts with and without screeds were taken from the latest engineering thesis [2], written under the supervision of the author of this monograph.

1. The calculation procedure for the theoretical maximum heat flux density of the first flooring covers layout with lamellas, depending on the distance between the heating pipes, with detailed technical data of the individual layers, is presented below:

- PVC vinyl floor panels, 5 mm thick, $\lambda = 0.45$ W/(m × K) [13],
- PU 560 polyurethane adhesive, 2 mm thick, $\lambda = 0.25$ W/(m × K) [14],
- BAUNORM Fibercement Board dry screed, 10 mm thick, $\lambda = 0.2166$ W/(m × K) [28],
- multilayer heating pipe with aluminium insert, d = 16 mm, s = 2 mm, $\lambda = 0.43$ W/(m × K) [33],
- aluminium heat dissipation lamellas, 0.09 mm thick, $\lambda = 200$ W/(m × K) [32],

Waterpanel X XPS insulation board, 10 mm thick, $\lambda = 0.2166$ W/(m×K), 30 mm, $\lambda = 0.034$ W/(m×K) or EPS Waterpanel E, thickness 30 mm, $\lambda = 0.036$ W/(m×K) [31].

$$q_i^{max} = 8.92\left(\bar{\theta}_F^{max} - \theta_i\right)^{1.1} \; [\mathrm{W/m^2}]$$

$$\bar{\theta}_F^{max} - \theta_i = k_{CL}\left(\theta_F^{max} - \theta_i\right) [°\mathrm{C}]$$

The thermal conductivity coefficient of the dissipating layer (lamella) k_{CL} for different coil spacings is

$$k_{CL,xps\,(10)} = \frac{L_U + 2L_{fin} \times k_{fin} + 0.01L_G \times k_{fin}^{min}}{T} = \frac{0.016 + 2 \times 0.0420 \times 0.880}{0.10} = 0.8989$$

$$k_{CL,xps\,(12.5)} = \frac{L_U + 2L_{fin} \times k_{fin} + 0.01L_G \times k_{fin}^{min}}{T} = \frac{0.016 + 2 \times 0.0545 \times 0.815}{0.125} = 0.8390$$

$$k_{CL,xps\,(15)} = \frac{L_U + 2L_{fin} \times k_{fin} + 0.01L_G \times k_{fin}^{min}}{T} = \frac{0.016 + 2 \times 0.0670 \times 0.749}{0.15} = 0.7759$$

$$k_{CL,xps\,(20)} = \frac{L_U + 2L_{fin} \times k_{fin} + 0.01L_G \times k_{fin}^{min}}{T} = \frac{0.016 + 2 \times 0.0920 \times 0.626}{0.20} = 0.6555$$

$$k_{CL,eps\,(10)} = \frac{L_U + 2L_{fin} \times k_{fin} + 0.01L_G \times k_{fin}^{min}}{T} = \frac{0.016 + 2 \times 0.042 \times 0.878}{0.10} = 0.8978$$

$$k_{CL,eps\,(12.5)} = \frac{L_U + 2L_{fin} \times k_{fin} + 0.01L_G \times k_{fin}^{min}}{T} = \frac{0.016 + 2 \times 0.0545 \times 0.814}{0.125} = 0.8376$$

$$k_{CL,eps\,(15)} = \frac{L_U + 2L_{fin} \times k_{fin} + 0.01L_G \times k_{fin}^{min}}{T} = \frac{0.016 + 2 \times 0.0670 \times 0.747}{0.15} = 0.7741$$

$$k_{CL,eps\,(20)} = \frac{L_U + 2L_{fin} \times k_{fin} + 0.01L_G \times k_{fin}^{min}}{T} = \frac{0.016 + 2 \times 0.0920 \times 0.623}{0.20} = 0.6533$$

$$L_{fin\,(10)} = \frac{T - L_U - L_G}{2} = \frac{0.10 - \; 0.016 - 0.00}{2} = 0.0420 \text{ m}$$

$$L_{fin\,(12.5)} = \frac{T - L_U - L_G}{2} = \frac{0.125 - 0.016 - 0.00}{2} = 0.0545 \text{ m}$$

$$L_{fin\,(15)} = \frac{T - L_U - L_G}{2} = \frac{0.15 - 0.016 - 0.00}{2} = 0.0670\text{ m}$$

$$L_{fin\,(20)} = \frac{T - L_U - L_G}{2} = \frac{0.20 - 0.016 - 0.00}{2} = 0.0920\text{ m}$$

$$k_{fin,xps\,(10)} = \frac{\ell}{L_{fin}} \tanh\left(\frac{L_{fin}}{\ell}\right) = \frac{0.0646}{0.0420} \tanh\left(\frac{0.0420}{0.0646}\right) = 0.880$$

$$k_{fin,xps\,(12.5)} = \frac{\ell}{L_{fin}} \tanh\left(\frac{L_{fin}}{\ell}\right) = \frac{0.0646}{0.0545} \tanh\left(\frac{0.0545}{0.0646}\right) = 0.815$$

$$k_{fin,xps\,(15)} = \frac{\ell}{L_{fin}} \tanh\left(\frac{L_{fin}}{\ell}\right) = \frac{0.0646}{0.0670} \tanh\left(\frac{0.0670}{0.0646}\right) = 0.749$$

$$k_{fin,xps\,(20)} = \frac{\ell}{L_{fin}} \tanh\left(\frac{L_{fin}}{\ell}\right) = \frac{0.0646}{0.0920} \tanh\left(\frac{0.0920}{0.0646}\right) = 0.626$$

$$k_{fin,eps\,(10)} = \frac{\ell}{L_{fin}} \tanh\left(\frac{L_{fin}}{\ell}\right) = \frac{0.0643}{0.0420} \tanh\left(\frac{0.0420}{0.0643}\right) = 0.878$$

$$k_{fin,eps\,(12.5)} = \frac{\ell}{L_{fin}} \tanh\left(\frac{L_{fin}}{\ell}\right) = \frac{0.0643}{0.0545} \tanh\left(\frac{0.0545}{0.0643}\right) = 0.814$$

$$k_{fin,eps\,(15)} = \frac{\ell}{L_{fin}} \tanh\left(\frac{L_{fin}}{\ell}\right) = \frac{0.0643}{0.0670} \tanh\left(\frac{0.0670}{0.0643}\right) = 0.747$$

$$k_{fin,eps\,(20)} = \frac{\ell}{L_{fin}} \tanh\left(\frac{L_{fin}}{\ell}\right) = \frac{0.0643}{0.0920} \tanh\left(\frac{0.0920}{0.0643}\right) = 0.623$$

$$\ell_{xps} = \sqrt{\frac{s_{WL} \times \lambda_{WL}}{\frac{1}{R_i} + \frac{1}{R_e}}} = \sqrt{\frac{0.00009 \times 200}{\frac{1}{0.308} + \frac{1}{0.941}}} = 0.0646$$

$$\ell_{eps} = \sqrt{\frac{s_{WL} \times \lambda_{WL}}{\frac{1}{R_i} + \frac{1}{R_e}}} = \sqrt{\frac{0.00009 \times 200}{\frac{1}{0.308} + \frac{1}{0.898}}} = 0.0643$$

where

$s_{WL} = 0.00009$ m – thickness of the conductive layer (aluminium lamellas)

$\lambda_{WL} = 200$ W/m×K – thermal conductivity of the conductive layer (aluminium)

The thermal resistance under the conductive layer to the room below or outside is

$$R_{e,xps\,(10)} = R_{e,0} - 2 \times \Delta R_e = 1.05 - 2 \times 0.0753 = 0.902 \ m^2K/W$$

$$R_{e,xps\,(12.5)} = R_{e,0} - 2 \times \Delta R_e = 1.05 - 2 \times 0.0602 = 0.932 \ m^2K/W$$

$$R_{e,xps\,(15)} = R_{e,0} - 2 \times \Delta R_e = 1.05 - 2 \times 0.0502 = 0.952 \ m^2K/W$$

$$R_{e,xps\,(20)} = R_{e,0} - 2 \times \Delta R_e = 1.05 - 2 \times 0.0376 = 0.977 \ m^2K/W$$

$$R_{e,eps\,(10)} = R_{e,0} - 2 \times \Delta R_e = 1.00 - 2 \times 0.0711 = 0.861 \ m^2K/W$$

$$R_{e,eps\,(12.5)} = R_{e,0} - 2 \times \Delta R_e = 1.00 - 2 \times 0.0569 = 0.890 \ m^2K/W$$

$$R_{e,eps\,(15)} = R_{e,0} - 2 \times \Delta R_e = 1.00 - 2 \times 0.0474 = 0.909 \ m^2K/W$$

$$R_{e,eps\,(20)} = R_{e,0} - 2 \times \Delta R_e = 1.00 - 2 \times 0.0356 = 0.932 \ m^2K/W$$

Small differences in the thermal resistance values of the layer transferring heat to the room below allowed for averaging. Therefore, the following values were adopted for further analysis:

$$R_{e,xps} = 0.941 \ m^2K/W$$

$$R_{e,eps} = 0.898 \ m^2K/W$$

$$\Delta R_{e,xps\,(10)} = \frac{d}{\lambda_{ins}} \times \frac{b}{T} = \frac{0.016}{0.034} \times \frac{0.016}{0.10} = 0.0753 \ m^2K/W$$

$$\Delta R_{e,xps\,(12.5)} = \frac{d}{\lambda_{ins}} \times \frac{b}{T} = \frac{0.016}{0.034} \times \frac{0.016}{0.125} = 0.0602 \ m^2K/W$$

$$\Delta R_{e,xps\,(15)} = \frac{d}{\lambda_{ins}} \times \frac{b}{T} = \frac{0.016}{0.034} \times \frac{0.016}{0.15} = 0.0502 \ m^2K/W$$

$$\Delta R_{e,xps\,(20)} = \frac{d}{\lambda_{ins}} \times \frac{b}{T} = \frac{0.016}{0.034} \times \frac{0.016}{0.20} = 0.0376 \ m^2K/W$$

$$\Delta R_{e,eps\,(10)} = \frac{d}{\lambda_{ins}} \times \frac{b}{T} = \frac{0.016}{0.036} \times \frac{0.016}{0.10} = 0.0711 \ m^2K/W$$

$$\Delta R_{e,eps\,(12.5)} = \frac{d}{\lambda_{ins}} \times \frac{b}{T} = \frac{0.016}{0.036} \times \frac{0.016}{0.125} = 0.0569\ m^2K/W$$

$$\Delta R_{e,eps\,(15)} = \frac{d}{\lambda_{ins}} \times \frac{b}{T} = \frac{0.016}{0.036} \times \frac{0.016}{0.15} = 0.0474\ m^2K/W$$

$$\Delta R_{e,eps\,(20)} = \frac{d}{\lambda_{ins}} \times \frac{b}{T} = \frac{0.016}{0.036} \times \frac{0.016}{0.20} = 0.0356\ m^2K/W$$

$$R_{e,0,xps} = R_u + R_{se} = \frac{d_k}{\lambda_k} + R_{se} = \frac{0.03}{0.034} + 0.17 = 1.05\ m^2K/W$$

$$R_{e,0,eps} = R_u + R_{se} = \frac{d_k}{\lambda_k} + R_{se} = \frac{0.03}{0.036} + 0.17 = 1.00\ m^2K/W$$

where

$R_{se} = 0.17\ m^2K/W$ – boundary layer resistance under the floor surface according to [5]

$$R_u = \Sigma_k \frac{d_k}{\lambda_k}$$

d_k – XPS or EPS thermal insulation thickness [m]

λ_k – thermal conductivity coefficient of thermal insulation [W/m × K]

The thermal resistance of the layer conducting heat to the heated room is

$$R_i = R_o + R_{con,i} + R_{si} = 0.065 + 0.15 + 0.093 = 0.308\ m^2K/W$$

The resistance for carefully shaped conductive layers permanently glued with an insulation layer is $R_{con,i} = 0.10\ m^2K/W$. Since the aluminium lamellas are not glued to the screed, a higher thermal resistance should be assumed for calculations, which is

$$R_{con,i} = 0.15\ m^2K/W$$

The thermal resistance of the floor layers above the conductive layer is

$$R_o = \frac{d_i}{\lambda_i} = \frac{0.01}{0.2166} + \frac{0.002}{0.25} + \frac{0.005}{0.45} = 0.046 + 0.008 + 0.011 = 0.065\ m^2K/W$$

The heat transfer resistance on the floor surface was assumed in accordance with the standard [5], as in the case of most underfloor heating systems, which is

$$R_{si} = \frac{1}{\alpha_i} = 0.093\ m^2K/W$$

The temperature difference between the maximum allowable average temperature and the air temperature, when we substitute k_{CL} values depending on the heating pipe spacing, XPS or EPS thermal insulation type, and the room's impact zone, is

- in the central zone

$$\overline{\theta}^{max}_{F,xps\,(10)} - \theta_{i,xps\,(10)} = k_{CL}\left(\theta^{max}_{F} - \theta_{i}\right) = 0.8989(29 - 20) = 8.0897°C$$

$$\overline{\theta}^{max}_{F,xps\,(12.5)} - \theta_{i,xps\,(12,5)} = k_{CL}\left(\theta^{max}_{F} - \theta_{i}\right) = 0.8390(29 - 20) = 7.5513°C$$

$$\overline{\theta}^{max}_{F,xps\,(15)} - \theta_{i,xps\,(15)} = k_{CL}\left(\theta^{max}_{F} - \theta_{i}\right) = 0.7759(29 - 20) = 6.9835°C$$

$$\overline{\theta}^{max}_{F,xps\,(20)} - \theta_{i,xps\,(20)} = k_{CL}\left(\theta^{max}_{F} - \theta_{i}\right) = 0.6555(29 - 20) = 5.8997°C$$

$$\overline{\theta}^{max}_{F,eps\,(10)} - \theta_{i,eps\,(10)} = k_{CL}\left(\theta^{max}_{F} - \theta_{i}\right) = 0.8978(29 - 20) = 8.0806°C$$

$$\overline{\theta}^{max}_{F,eps\,(12.5)} - \theta_{i,eps\,(12,5)} = k_{CL}\left(\theta^{max}_{F} - \theta_{i}\right) = 0.8376(29 - 20) = 7.5380°C$$

$$\overline{\theta}^{max}_{F,eps\,(15)} - \theta_{i,eps\,(15)} = k_{CL}\left(\theta^{max}_{F} - \theta_{i}\right) = 0.7741(29 - 20) = 6.9669°C$$

$$\overline{\theta}^{max}_{F,eps\,(20)} - \theta_{i,eps\,(20)} = k_{CL}\left(\theta^{max}_{F} - \theta_{i}\right) = 0.6533(29 - 20) = 5.8794°C$$

- in the edge zone

$$\overline{\theta}^{max}_{F,xps\,(10)} - \theta_{i,xps\,(10)} = k_{CL}\left(\theta^{max}_{F} - \theta_{i}\right) = 0.8989(35 - 20) = 13.4829°C$$

$$\overline{\theta}^{max}_{F,xps\,(12.5)} - \theta_{i,xps\,(12,5)} = k_{CL}\left(\theta^{max}_{F} - \theta_{i}\right) = 0.8390(35 - 20) = 12.5855°C$$

$$\overline{\theta}^{max}_{F,xps\,(15)} - \theta_{i,xps\,(15)} = k_{CL}\left(\theta^{max}_{F} - \theta_{i}\right) = 0.7759(35 - 20) = 11.6392°C$$

$$\overline{\theta}^{max}_{F,xps\,(20)} - \theta_{i,xps\,(20)} = k_{CL}\left(\theta^{max}_{F} - \theta_{i}\right) = 0.6555(35 - 20) = 9.8328°C$$

$$\overline{\theta}^{max}_{F,eps\,(10)} - \theta_{i,eps\,(10)} = k_{CL}\left(\theta^{max}_{F} - \theta_{i}\right) = 0.8978(35 - 20) = 13.4676°C$$

$$\overline{\theta}^{max}_{F,eps\,(12.5)} - \theta_{i,eps\,(12,5)} = k_{CL}\left(\theta^{max}_{F} - \theta_{i}\right) = 0.8376(35 - 20) = 12.5634°C$$

$$\overline{\theta}^{max}_{F,eps\,(15)} - \theta_{i,eps\,(15)} = k_{CL}\left(\theta^{max}_{F} - \theta_{i}\right) = 0.7741(35 - 20) = 11.6116°C$$

$$\overline{\theta}^{max}_{F,eps\,(20)} - \theta_{i,eps\,(20)} = k_{CL}\left(\theta^{max}_{F} - \theta_{i}\right) = 0.6533(35 - 20) = 9.7991°C$$

Ultimately, the calculated maximum heat flux density, depending on the coil spacing and the type of EPS or XPS thermal insulation, assuming an indoor temperature of 20°C and an impact zone of q_i^{max} is

- in the central zone

$$q_{xps\,(10)}^{max} = 8.92\left(\overline{\theta}_F^{max} - \theta_i\right)^{1.1} = 8.92(8.0897)^{1.1} = 88.94\ W/m^2$$

$$q_{xps\,(12.5)}^{max} = 8.92\left(\overline{\theta}_F^{max} - \theta_i\right)^{1.1} = 8.92(7.5513)^{1.1} = 82.45\ W/m^2$$

$$q_{xps\,(15)}^{max} = 8.92\left(\overline{\theta}_F^{max} - \theta_i\right)^{1.1} = 8.92(6.9835)^{1.1} = 75.66\ W/m^2$$

$$q_{xps\,(20)}^{max} = 8.92\left(\overline{\theta}_F^{max} - \theta_i\right)^{1.1} = 8.92(5.8997)^{1.1} = 62.85\ W/m^2$$

$$q_{eps\,(10)}^{max} = 8.92\left(\overline{\theta}_F^{max} - \theta_i\right)^{1.1} = 8.92(8.0806)^{1.1} = 88.83\ W/m^2$$

$$q_{eps\,(12.5)}^{max} = 8.92\left(\overline{\theta}_F^{max} - \theta_i\right)^{1.1} = 8.92(7.5380)^{1.1} = 82.29\ W/m^2$$

$$q_{eps\,(15)}^{max} = 8.92\left(\overline{\theta}_F^{max} - \theta_i\right)^{1.1} = 8.92(6.9669)^{1.1} = 75.46\ W/m^2$$

$$q_{eps\,(20)}^{max} = 8.92\left(\overline{\theta}_F^{max} - \theta_i\right)^{1.1} = 8.92(5.8794)^{1.1} = 62.61\ W/m^2$$

- in the edge zone

$$q_{xps\,(10)}^{max} = 8.92\left(\overline{\theta}_F^{max} - \theta_i\right)^{1.1} = 8.92(13.4829)^{1.1} = 156.00\ W/m^2$$

$$q_{xps\,(12.5)}^{max} = 8.92\left(\overline{\theta}_F^{max} - \theta_i\right)^{1.1} = 8.92(12.5855)^{1.1} = 144.62\ W/m^2$$

$$q_{xps\,(15)}^{max} = 8.92\left(\overline{\theta}_F^{max} - \theta_i\right)^{1.1} = 8.92(11.6392)^{1.1} = 132.70\ W/m^2$$

$$q_{xps\,(20)}^{max} = 8.92\left(\overline{\theta}_F^{max} - \theta_i\right)^{1.1} = 8.92(9.8328)^{1.1} = 110.23\ W/m^2$$

$$q_{eps\,(10)}^{max} = 8.92\left(\overline{\theta}_F^{max} - \theta_i\right)^{1.1} = 8.92(13.4676)^{1.1} = 155.81\ W/m^2$$

$$q^{max}_{eps\,(12.5)} = 8.92\left(\bar{\theta}^{max}_{F} - \theta_i\right)^{1.1} = 8.92(12.5634)^{1.1} = 144.34\ W/m^2$$

$$q^{max}_{eps\,(15)} = 8.92\left(\bar{\theta}^{max}_{F} - \theta_i\right)^{1.1} = 8.92(11.6116)^{1.1} = 132.36\ W/m^2$$

$$q^{max}_{eps\,(20)} = 8.92\left(\bar{\theta}^{max}_{F} - \theta_i\right)^{1.1} = 8.92(9.7991)^{1.1} = 109.82\ W/m^2$$

By substituting the values calculated above, we calculate the balanced heat transfer coefficient K_H, depending on the thermal insulation used and the distance of the heating pipes,

$$K_{H,xps\,(10)} = \frac{q_i^{max}}{\theta_F^{max} - \theta_i} = \frac{88.94}{29 - 20} = 9.822\ W/m^2K$$

$$K_{H,xps\,(12.5)} = \frac{q_i^{max}}{\theta_F^{max} - \theta_i} = \frac{82.45}{29 - 20} = 9.161\ W/m^2K$$

$$K_{H,xps\,(15)} = \frac{q_i^{max}}{\theta_F^{max} - \theta_i} = \frac{75.66}{29 - 20} = 8.406\ W/m^2K$$

$$K_{H,xps\,(20)} = \frac{q_i^{max}}{\theta_F^{max} - \theta_i} = \frac{62.85}{29 - 20} = 6.983\ W/m^2K$$

$$K_{H,eps\,(10)} = \frac{q_i^{max}}{\theta_F^{max} - \theta_i} = \frac{88.83}{29 - 20} = 9.870\ W/m^2K$$

$$K_{H,eps\,(12.5)} = \frac{q_i^{max}}{\theta_F^{max} - \theta_i} = \frac{82.29}{29 - 20} = 9.143\ W/m^2K$$

$$K_{H,eps\,(15)} = \frac{q_i^{max}}{\theta_F^{max} - \theta_i} = \frac{75.46}{29 - 20} = 8.384\ W/m^2K$$

$$K_{H,eps\,(20)} = \frac{q_i^{max}}{\theta_F^{max} - \theta_i} = \frac{62.61}{29 - 20} = 6.957\ W/m^2K$$

The final value of the calculated heat flux density q_{ol}, emitted from the first heated lightweight floor layout with dry screed and aluminium lamellas, taking into account the assumed criteria of the adopted example variant of layers with vinyl panel flooring at different coil spacings, is

$$q_{ol} = K_H \times \Delta\theta_H\ [W/m^2]$$

$$\Delta\theta_H = \theta_H - \theta_i$$

a. with a coil spacing of 10 cm

$$q_{xps,10\,(28^\circ)} = 9.882 \times (28 - 20) = 79.06\ \mathrm{W/m^2}$$

$$q_{xps,10\,(35^\circ)} = 9.882 \times (35 - 20) = 148.23\ \mathrm{W/m^2}$$

$$q_{xps,10\,(40^\circ)} = 9.882 \times (40 - 20) = 197.64\ \mathrm{W/m^2}$$

$$q_{xps,10\,(45^\circ)} = 9.882 \times (45 - 20) = 247.05\ \mathrm{W/m^2}$$

$$q_{eps,10\,(28^\circ)} = 9.870 \times (28 - 20) = 78.96\ \mathrm{W/m^2}$$

$$q_{eps,10\,(35^\circ)} = 9.870 \times (35 - 20) = 148.05\ \mathrm{W/m^2}$$

$$q_{eps,10\,(40^\circ)} = 9.870 \times (40 - 20) = 197.40\ \mathrm{W/m^2}$$

$$q_{eps,10\,(45^\circ)} = 9.870 \times (45 - 20) = 246.75\ \mathrm{W/m^2}$$

b. with a coil spacing of 12.5 cm

$$q_{xps,12.5\,(28^\circ)} = 9.161 \times (28 - 20) = 73.29\ \mathrm{W/m^2}$$

$$q_{xps,12.5\,(35^\circ)} = 9.161 \times (35 - 20) = 137.42\ \mathrm{W/m^2}$$

$$q_{xps,12.5\,(40^\circ)} = 9.161 \times (40 - 20) = 183.22\ \mathrm{W/m^2}$$

$$q_{xps,12.5\,(45^\circ)} = 9.161 \times (45 - 20) = 229.03\ \mathrm{W/m^2}$$

$$q_{eps,12.5\,(28^\circ)} = 9.143 \times (28 - 20) = 73.15\ \mathrm{W/m^2}$$

$$q_{eps,12.5\,(35^\circ)} = 9.143 \times (35 - 20) = 137.15\ \mathrm{W/m^2}$$

$$q_{eps,12.5\,(40^\circ)} = 9.143 \times (40 - 20) = 182.87\ \mathrm{W/m^2}$$

$$q_{eps,12.5\,(45^\circ)} = 9.143 \times (45 - 20) = 228.58\ \mathrm{W/m^2}$$

c. with a coil spacing of 15 cm

$$q_{xps,15\ (28°)} = 8.406 \times (28 - 20) = 67.25\ W/m^2$$

$$q_{xps,15\ (35°)} = 8.406 \times (35 - 20) = 126.09\ W/m^2$$

$$q_{xps,15\ (40°)} = 8.406 \times (40 - 20) = 168.13\ W/m^2$$

$$q_{xps,15\ (45°)} = 8.406 \times (45 - 20) = 210.16\ W/m^2$$

$$q_{eps,15\ (28°)} = 8.384 \times (28 - 20) = 67.07\ W/m^2$$

$$q_{eps,15\ (35°)} = 8.384 \times (35 - 20) = 125.77\ W/m^2$$

$$q_{eps,15\ (40°)} = 8.384 \times (40 - 20) = 167.69\ W/m^2$$

$$q_{eps,15\ (45°)} = 8.384 \times (45 - 20) = 209.61\ W/m^2$$

d. with a coil spacing of 20 cm

$$q_{xps,20\ (28°)} = 6.983 \times (28 - 20) = 55.86\ W/m^2$$

$$q_{xps,20\ (35°)} = 6.983 \times (35 - 20) = 104.74\ W/m^2$$

$$q_{xps,20\ (40°)} = 6.983 \times (40 - 20) = 139.66\ W/m^2$$

$$q_{xps,20\ (45°)} = 6.983 \times (45 - 20) = 174.57\ W/m^2$$

$$q_{eps,20\ (28°)} = 6.957 \times (28 - 20) = 55.65\ W/m^2$$

$$q_{eps,20\ (35°)} = 6.957 \times (35 - 20) = 104.35\ W/m^2$$

$$q_{eps,20\ (40°)} = 6.957 \times (40 - 20) = 139.13\ W/m^2$$

$$q_{eps,20\ (45°)} = 6.957 \times (45 - 20) = 173.91\ W/m^2$$

2. The maximum permissible temperature of the medium in the heating pipe of the first exemplary variant of the heated floor layers covered with vinyl panels with dry screed and metal lamellas θ_H^{max} is

$$\theta_H^{max} = \theta_i + q_i^{max}\left(R_i + \frac{1}{\eta} \times R_{HC}\right)$$

$$\theta_{H,xps,10}^{max} = 20 + 88.94\left(0.308 + \frac{1}{0.753} \times 0.0383\right) = 51.95°C$$

$$\theta_{H,xps,12.5}^{max} = 20 + 82.45\left(0.308 + \frac{1}{0.753} \times 0.0603\right) = 52.02°C$$

$$\theta_{H,xps,15}^{max} = 20 + 75.66\left(0.308 + \frac{1}{0.753} \times 0.0865\right) = 52.02°C$$

$$\theta_{H,xps,20}^{max} = 20 + 62.85\left(0.308 + \frac{1}{0.753} \times 0.1496\right) = 51.85°C$$

$$\theta_{H,eps,10}^{max} = 20 + 88.83\left(0.308 + \frac{1}{0.744} \times 0.0383\right) = 51.93°C$$

$$\theta_{H,eps,12.5}^{max} = 20 + 82.29\left(0.308 + \frac{1}{0.744} \times 0.0603\right) = 52.01°C$$

$$\theta_{H,eps,15}^{max} = 20 + 75.46\left(0.308 + \frac{1}{0.744} \times 0.0865\right) = 52.01°C$$

$$\theta_{H,eps,20}^{max} = 20 + 62.61\left(0.308 + \frac{1}{0.744} \times 0.1493\right) = 51.85°C$$

The calculations for the individual components required to determine the maximum permissible medium temperature in the heating pipe θ_H^{max}, given the assumed heating pipe spacing and EPS and XPS thermal insulation, are provided below.

The efficiency of the underfloor heating system is as follows:

$$\eta_{xps} = \frac{R_{e,xps}}{R_i + R_{e,xps}} = \frac{0.941}{0.308 + 0.941} = 0.753$$

$$\eta_{eps} = \frac{R_{e,eps}}{R_i + R_{e,eps}} = \frac{0.898}{0.308 + 0.898} = 0.744$$

The total thermal resistance between the heat source and the conductive layer for four coil spacings and EPS and XPS thermal insulation is

$$R_{HC} = T \times R'_R + T \times R'_{R,con} \frac{T}{2} \times R'_U + R_{CL} \; [m^2 \times °C / W]$$

$$R_{HC,xps,10} = 0.10 \times 0.107 + 0.10 \times 0.625 \times \frac{0.10}{2} \times 0.5 + 0.0261 = 0.0383 \; m^2 \times °C/ W$$

$$R_{HC,xps,12.5} = 0.125 \times 0.107 + 0.125 \times 0.625 \times \frac{0.125}{2} \times 0.5 + 0.0445 = 0.0603 \; m^2 \times °C/ W$$

$$R_{HC,xps,15} = 0.15 \times 0.107 + 0.15 \times 0.625 \times \frac{0.15}{2} \times 0.5 + 0.0670 = 0.0865 \; m^2 \times °C/ W$$

$$R_{HC,xps,20} = 0.20 \times 0.107 + 0.20 \times 0.625 \times \frac{0.20}{2} \times 0.5 + 0.1220 = 0.1496 \; m^2 \times °C/ W$$

$$R_{HC,eps,10} = 0.10 \times 0.107 + 0.10 \times 0.625 \times \frac{0.10}{2} \times 0.5 + 0.0261 = 0.0383 \; m^2 \times °C/ W$$

$$R_{HC,eps,12.5} = 0.125 \times 0.107 + 0.125 \times 0.625 \times \frac{0.125}{2} \times 0.5 + 0.0445 = 0.0603 \; m^2 \times °C/ W$$

$$R_{HC,eps,15} = 0.15 \times 0.107 + 0.15 \times 0.625 \times \frac{0.15}{2} \times 0.5 + 0.0670 = 0.0865 \; m^2 \times °C/ W$$

$$R_{HC,eps,20} = 0.20 \times 0.107 + 0.20 \times 0.625 \times \frac{0.20}{2} \times 0.5 + 0.1218 = 0.1493 \; m^2 \times °C/ W$$

where

$$R'_U = \frac{0.008}{d_a} = \frac{0.008}{0.016} = 0.50 \; m^2 \times °C / W$$

$$R'_{R,con} = \frac{0.01}{d_a} = \frac{0.01}{0.016} = 0.625 \; m^2 \times °C / W$$

$$R'_R = \frac{1}{2\pi\lambda_R} \ln\left(\frac{d_o}{d_{ip}}\right) = \frac{1}{2\pi \times 0.43} \ln\left(\frac{0.016}{0.012}\right) = 0.107 \; m^2 \times °C / W$$

The fictitious thermal resistance of the conductive layer for EPS and XPS thermal insulation at different coil spacings is

$$R_{CL,xps,10} = \frac{1}{\frac{1}{R_i} + \frac{1}{R_e}}\left(\frac{1}{k_{CL}} - 1\right) = \frac{1}{\frac{1}{0.308} + \frac{1}{0.941}}\left(\frac{1}{0.8989} - 1\right) = 0.0261\ m^2 \times \frac{°C}{W}$$

$$R_{CL,xps,12.5} = \frac{1}{\frac{1}{R_i} + \frac{1}{R_e}}\left(\frac{1}{k_{CL}} - 1\right) = \frac{1}{\frac{1}{0.308} + \frac{1}{0.941}}\left(\frac{1}{0.8390} - 1\right) = 0.0445\ m^2 \times \frac{°C}{W}$$

$$R_{CL,xps,15} = \frac{1}{\frac{1}{R_i} + \frac{1}{R_e}}\left(\frac{1}{k_{CL}} - 1\right) = \frac{1}{\frac{1}{0.308} + \frac{1}{0.941}}\left(\frac{1}{0.7759} - 1\right) = 0.0670\ m^2 \times \frac{°C}{W}$$

$$R_{CL,xps,20} = \frac{1}{\frac{1}{R_i} + \frac{1}{R_e}}\left(\frac{1}{k_{CL}} - 1\right) = \frac{1}{\frac{1}{0.308} + \frac{1}{0.941}}\left(\frac{1}{0.6555} - 1\right) = 0.1220\ m^2 \times \frac{°C}{W}$$

$$R_{CL,eps,10} = \frac{1}{\frac{1}{R_i} + \frac{1}{R_e}}\left(\frac{1}{k_{CL}} - 1\right) = \frac{1}{\frac{1}{0.308} + \frac{1}{0.898}}\left(\frac{1}{0.8978} - 1\right) = 0.0261\ m^2 \times \frac{°C}{W}$$

$$R_{CL,eps,12.5} = \frac{1}{\frac{1}{R_i} + \frac{1}{R_e}}\left(\frac{1}{k_{CL}} - 1\right) = \frac{1}{\frac{1}{0.308} + \frac{1}{0.898}}\left(\frac{1}{0.8376} - 1\right) = 0.0445\ m^2 \times \frac{°C}{W}$$

$$R_{CL,eps,15} = \frac{1}{\frac{1}{R_i} + \frac{1}{R_e}}\left(\frac{1}{k_{CL}} - 1\right) = \frac{1}{\frac{1}{0.308} + \frac{1}{0.898}}\left(\frac{1}{0.7741} - 1\right) = 0.0670\ m^2 \times \frac{°C}{W}$$

$$R_{CL,eps,20} = \frac{1}{\frac{1}{R_i} + \frac{1}{R_e}}\left(\frac{1}{k_{CL}} - 1\right) = \frac{1}{\frac{1}{0.308} + \frac{1}{0.898}}\left(\frac{1}{0.6533} - 1\right) = 0.1218\ m^2 \times \frac{°C}{W}$$

Based on the calculations of the maximum permissible temperature of the medium in the heating pipe of the first exemplary variant of the heated floor layers covered with vinyl panels with dry screed and metal lamellas θ_H^{max}, it should be assumed that its safe use value should not exceed 51°C.

The calculation example for the first heated floor layout variants with vinyl flooring, BAUNORM FIBERCEMENT dry screed, and aluminium lamellas provides the basis for the calculations of the remaining first-layer layout variants. The set of mathematical formulas adopted in the example variant was applied to the calculations of other heated floor variants with KNAUF BRIO and KNAUF DUAL FLOOR dry screeds, as well as other floorings. The following seven flooring types were assumed for underfloor heating.

- PVC flooring with integrated XPO underlay
- three-layer wooden board + polyurethane adhesive

- laminate panels + smart underlay
- two-component microcement Microflex
- PVC flooring + Eurocol 640 Eurostar adhesive
- carpet flooring + Eurocol 540 Eurosafe Special adhesive
- flexible resin flooring

The results of the thermal power calculations for the first heated lightweight floor layout with thin screeds and aluminium lamellas are summarised in Table 5.3.

TABLE 5.3

Design Values of Thermal Power for the Eight Flooring Covers of the First Heated Floor Layout with Screeds and Aluminium Lamellas [2]

Coil Spacing [cm]		Design Thermal Efficiency, q_{ol} [W/m^2]											
		KNAUF BRIO				BAUNORM FIBERCEMENT				KNAUF DUAL FLOOR			
		Coil Supply Temperature [°C]											
		28	35	40	45	28	35	40	45	28	35	40	45
PVC Floor Panels – Vinyl + Polyurethane Adhesive PU 560													
XPS	10	79.55	149.15	198.87	248.59	79.06	148.23	197.64	247.05	80.01	150.02	200.03	250.03
	12.5	74.00	138.75	185.00	231.25	73.29	137.42	183.22	229.03	74.67	140.01	186.69	233.36
	15	68.14	127.76	170.35	212.94	67.25	126.09	168.13	210.16	68.99	129.35	172.47	215.59
	20	56.94	106.77	142.35	177.94	55.86	104.74	148.05	174.57	57.99	108.73	144.97	181.22
EPS	10	79.45	148.97	198.62	248.28	78.96	148.05	197.40	246.75	79.91	149.83	199.77	249.72
	12.5	73.85	138.48	184.64	230.80	73.15	137.15	182.87	228.58	74.53	139.73	186.31	232.89
	15	67.96	127.42	169.89	212.37	67.07	125.77	167.69	209.61	68.80	129.00	172.00	215.00
	20	56.72	106.35	141.80	177.25	55.65	104.35	139.13	173.91	57.76	108.29	144.39	180.48
PVC Floor Panels – Vinyl with Integrated XPO Underlay													
XPS	10	79.71	149.46	199.28	249.10	79.25	148.59	198.11	247.64	80.15	150.29	200.38	250.48
	12.5	74.24	139.20	185.60	232.00	73.56	137.93	183.90	229.88	74.88	140.40	187.21	234.01
	15	68.44	128.32	171.10	213.87	67.59	126.73	168.97	211.22	69.25	129.85	173.13	216.41
	20	57.31	107.45	143.27	179.09	56.27	105.51	140.68	175.85	58.32	109.35	145.80	182.24
EPS	10	79.61	149.27	199.03	248.79	79.15	148.40	197.87	247.33	80.05	150.09	200.13	250.16
	12.5	74.09	138.92	185.23	231.54	73.42	137.66	183.54	229.43	74.73	140.12	186.83	233.54
	15	68.25	127.98	170.64	213.29	67.41	126.40	168.53	210.66	69.06	129.49	172.66	215.82
	20	57.08	107.03	142.70	178.38	56.06	105.11	140.14	175.18	58.08	108.90	145.20	181.50
Three-Layer Wooden Board + Polyurethane Adhesive PU 560													
XPS	10	80.47	150.88	201.17	251.47	80.11	150.20	200.27	250.34	80.82	151.53	202.04	252.55
	12.5	75.35	141.28	188.37	235.47	74.82	140.28	187.04	233.81	75.86	142.24	189.66	237.07
	15	69.85	130.96	174.62	218.27	69.17	129.70	172.93	216.16	70.50	132.20	176.26	220.33
	20	59.07	110.75	147.67	184.58	58.22	109.15	145.54	181.92	59.90	112.32	149.76	187.20
EPS	10	80.37	150.69	200.92	251.14	80.01	150.01	200.02	250.02	80.71	151.33	201.78	252.22
	12.5	75.20	140.99	187.99	234.99	74.67	140.00	186.67	233.34	75.71	141.95	189.27	236.58
	15	69.65	130.60	174.13	217.67	68.98	129.34	172.45	215.57	70.31	131.82	175.76	219.70
	20	58.82	110.29	147.05	183.81	57.98	108.71	144.95	181.19	59.65	111.84	149.12	186.40

(Continued)

TABLE 5.3 (*Continued*)
Design Values of Thermal Power for the Eight Flooring Covers of the First Heated Floor Layout with Screeds and Aluminium Lamellas [2]

Coil Spacing [cm]		Design Thermal Efficiency, q_{ol} [W/m^2]											
		KNAUF BRIO				BAUNORM FIBERCEMENT				KNAUF DUAL FLOOR			
		Coil Supply Temperature [°C]											
		28	35	40	45	28	35	40	45	28	35	40	45
		Laminate Flooring + Smart Underlay											
XPS	10	81.25	152.34	203.12	253.90	80.98	151.84	202.46	253.07	81.51	152.83	203.77	254.72
	12.5	76.51	143.46	191.28	239.10	76.11	142.71	190.28	237.85	76.90	144.19	192.26	240.32
	15	71.34	133.77	178.36	222.95	70.83	132.80	177.07	221.34	71.85	134.72	179.63	224.54
	20	60.99	114.35	152.47	190.59	60.32	113.10	150.80	188.50	61.65	115.60	154.13	192.67
EPS	10	81.14	152.14	202.86	253.57	80.88	151.65	202.19	252.74	81.40	152.63	203.50	254.38
	12.5	76.35	143.16	190.88	238.60	75.96	142.42	189.89	237.36	76.74	143.89	191.85	239.81
	15	71.14	133.38	177.84	222.31	70.63	132.42	176.56	220.70	71.64	134.33	179.10	223.88
	20	60.72	113.85	151.80	189.75	60.06	112.61	150.15	187.68	61.38	115.08	153.44	191.80
		Microflex Two-Component Microcement											
XPS	10	79.26	148.62	198.16	247.70	78.73	147.61	196.81	246.02	79.76	149.56	199.41	249.26
	12.5	73.59	137.97	183.96	229.95	72.81	136.52	182.03	227.54	74.31	139.34	185.78	232.23
	15	67.62	126.79	169.05	211.31	66.66	124.99	166.65	208.32	68.53	128.50	171.33	214.17
	20	56.31	105.58	140.77	175.97	55.16	103.42	137.90	172.37	57.42	107.67	143.56	179.45
EPS	10	79.16	148.43	197.91	247.39	78.63	147.43	196.57	245.71	79.66	149.37	199.16	248.95
	12.5	73.44	137.70	183.61	229.51	72.67	136.26	181.68	227.11	74.17	139.06	185.42	231.77
	15	67.44	126.45	168.60	210.76	66.49	124.67	166.22	207.78	68.35	128.15	170.87	213.59
	20	56.09	105.18	140.23	175.29	54.95	103.04	137.39	171.73	57.20	107.24	142.99	178.74
		PVC Flooring + Eurocol 640 Eurostar Special Adhesive											
XPS	10	79.49	149.05	198.73	248.42	78.99	148.11	197.48	246.85	79.96	149.93	199.91	249.88
	12.5	73.92	138.60	184.80	231.00	73.20	137.24	182.99	228.74	74.60	139.88	186.51	233.14
	15	68.04	127.57	170.10	212.62	67.14	125.88	167.84	209.80	68.90	129.19	172.25	215.31
	20	56.82	106.53	142.04	177.55	55.73	104.48	139.31	174.14	57.88	108.52	144.70	180.87
EPS	10	79.39	148.86	198.48	248.10	78.89	147.93	197.24	246.55	79.86	149.74	199.65	249.57
	12.5	73.77	138.33	184.44	230.55	73.06	136.98	182.64	228.30	74.46	139.60	186.14	232.67
	15	67.86	127.23	169.64	212.05	66.96	125.55	167.40	209.25	68.71	128.83	171.78	214.72
	20	56.60	106.12	141.49	176.86	55.52	104.09	138.79	173.49	57.65	108.09	144.11	180.14
		Flotex Colour Carpet + Eurocol 640 Eurosafe Special Adhesive											
XPS	10	80.47	150.89	201.18	251.48	80.11	150.21	200.28	250.35	80.82	151.53	202.04	252.56
	12.5	75.35	141.29	188.39	235.48	74.82	140.29	187.06	233.82	75.87	142.25	189.67	237.08
	15	69.85	130.98	174.63	218.29	69.18	129.71	172.94	216.18	70.51	132.21	176.28	220.34
	20	59.07	110.76	147.69	184.61	58.22	109.17	145.56	181.95	59.91	112.33	149.78	187.22
EPS	10	80.37	150.69	200.92	251.15	80.01	150.02	200.03	250.03	80.71	151.34	201.78	252.23
	12.5	75.20	141.00	188.00	235.01	74.67	140.01	186.68	233.35	75.71	141.96	189.28	236.60
	15	69.66	130.61	174.15	217.68	68.99	129.35	172.47	215.59	70.31	131.83	175.78	219.72
	20	58.83	110.30	147.07	183.84	57.99	108.73	144.97	181.21	59.65	111.85	149.14	186.42

(*Continued*)

TABLE 5.3 (*Continued*)
Design Values of Thermal Power for the Eight Flooring Covers of the First Heated Floor Layout with Screeds and Aluminium Lamellas [2]

Coil Spacing [cm]		Design Thermal Efficiency, q_{ol} [W/m^2]											
		KNAUF BRIO				BAUNORM FIBERCEMENT				KNAUF DUAL FLOOR			
		Coil Supply Temperature [°C]											
		28	35	40	45	28	35	40	45	28	35	40	45
		Sikafloor Elastic Resin Floor – 300 N + Sikafloor 156 Primer											
XPS	10	79.48	149.02	198.70	248.37	78.97	148.08	197.44	246.80	79.95	149.91	199.87	249.84
	12.5	73.90	138.56	184.74	230.93	73.17	137.19	182.93	228.66	74.58	139.85	186.46	233.08
	15	68.01	127.52	170.03	212.53	67.10	125.82	167.76	209.70	68.87	129.14	172.19	215.23
	20	56.78	106.47	141.96	177.45	55.69	104.41	139.22	174.02	57.85	108.46	144.62	180.77
EPS	10	79.38	148.83	198.45	248.06	78.88	147.89	197.19	246.49	79.85	149.71	199.62	249.52
	12.5	73.75	138.29	184.38	230.48	73.03	136.93	182.57	228.22	74.44	139.57	186.09	232.61
	15	67.83	127.18	169.57	211.96	66.93	125.49	167.32	209.15	68.69	128.79	171.72	214.65
	20	56.56	106.05	141.41	176.76	55.48	104.02	138.69	173.37	57.62	108.03	144.04	180.05

5.1.3 Determination of the Design Thermal Power of an Underfloor Heating Layout with Thin Screeds *without Lamellas*

The second heated floor layout, with thin screeds but without metal lamellas, was subjected to mathematical calculations using the formulas from [34]. This subsection presents an example calculation of the heat flux density (thermal power) for a specific configuration of selected layers from the second layout. Similar to the lamellas layout, the example calculations assumed a floor made of glued vinyl panels. It is bonded to a 10-mm-thin screed made of BAUNORM Fibercement Board fibre-cement boards using a polyurethane adhesive designed by the vinyl panel manufacturer. The remaining material assumptions, technical guidelines, and thermal power calculation variants were not changed but were based on the formulas developed in [34]. This allows comparison of the final calculated thermal power values for multi-variant selections of different types of flooring and dry screeds, with and without heat-dissipating metal lamellas. In the latter layout of heated floors without lamellas, the maximum water temperature in the coil was not determined. Currently, there is no reliable mathematical procedure for calculating this maximum temperature in a lightweight heated floor system, and the calculation procedure based on the Scandinavian standard [5] is not applicable. As in the first layout, all the components of the formulas given below were already listed and described in Chapter 4, which concerns the calculation of the thermal power of a floor without slats, so we have not attempted to redescribe their meaning in detail. As with the calculation of a floor with slats, the example of heat flux density calculations for a heated floor without screeds was taken from the latest engineering thesis [2], written under the supervision of the author of this monograph. It is important to note that in the second layout, the heat-conducting layer was a dry screed that also ensured the floor structure's stiffness and stability.

1. The calculation procedure for the theoretical maximum heat flux density of the second flooring covers layout **without lamellas**, depending on the distance of the heating pipes, with detailed technical data of the individual layers, is presented below:
   - PVC vinyl floor panels, 5 mm thick, $\lambda = 0.45$ W/(m × K) [13],
   - PU 560 polyurethane adhesive, 2 mm thick, $\lambda = 0.25$ W/(m × K) [14],
   - BAUNORM Fibercement Board dry screed, 10 mm thick, $\lambda = 0.2166$ W/(m × K) [28],
   - multilayer heating pipe with aluminium insert, d = 16 mm, s = 2 mm, $\lambda = 0.43$ W/(m × K) [33],
   - Waterpanel X XPS insulation board, 10 mm thick, $\lambda = 0.2166$ W/(m × K), 30 mm, $\lambda = 0.034$ W/(m × K) or EPS Waterpanel E, thickness 30 mm, $\lambda = 0.036$ W/(m × K) [31].

$$q_B^{max} = 8.92\left(\bar{\theta}_F^{max} - \theta_i\right)^{1.1} \text{ [W/m}^2\text{]}$$

$$\bar{\theta}_F^{max} - \theta_i = k_{CB}\left(\theta_F^{max} - \theta_i\right)$$

The thermal conductivity coefficient of the dissipating layer (lamella) k_{CB}, for different coil spacings is

$$k_{CB,xps\,(10)} = \frac{L_U + 2L_k \times k_k}{T} = \frac{0.016 + 2 \times 0.042 \times 0.484}{0.10} = 0.566 \text{ W/m}^2\text{K}$$

$$k_{CB,xps\,(12.5)} = \frac{L_U + 2L_k \times k_k}{T} = \frac{0.016 + 2 \times 0.0545 \times 0.382}{0.125} = 0.4613 \text{ W/m}^2\text{K}$$

$$k_{CB,xps\,(15)} = \frac{L_U + 2L_k \times k_k}{T} = \frac{0.016 + 2 \times 0.067 \times 0.313}{0.15} = 0.3867 \text{ W/m}^2\text{K}$$

$$k_{CB,xps\,(20)} = \frac{L_U + 2L_k \times k_k}{T} = \frac{0.016 + 2 \times 0.092 \times 0.229}{0.20} = 0.2906 \text{ W/m}^2\text{K}$$

$$k_{CB,eps\,(10)} = \frac{L_U + 2L_k \times k_k}{T} = \frac{0.016 + 2 \times 0.042 \times 0.481}{0.10} = 0.5643 \text{ W/m}^2\text{K}$$

$$k_{CB,eps\,(12.5)} = \frac{L_U + 2L_k \times k_k}{T} = \frac{0.016 + 2 \times 0.0545 \times 0.380}{0.125} = 0.4597 \text{ W/m}^2\text{K}$$

$$k_{CB,eps\,(15)} = \frac{L_U + 2L_k \times k_k}{T} = \frac{0.016 + 2 \times 0.067 \times 0.312}{0.15} = 0.3852\ \text{W/m}^2\text{K}$$

$$k_{CB,eps\,(20)} = \frac{L_U + 2L_k \times k_k}{T} = \frac{0.016 + 2 \times 0.092 \times 0.228}{0.20} = 0.2896\ \text{W/m}^2\text{K}$$

An integral element of the formula for the thermal conductivity coefficient of the adhesive layer and the floor without lamellas k_{CB} is the calculation below for different coil spacings and assuming no gap between the adhesive and the flooring:

$$L_{k\,(10)} = \frac{T - L_U}{2} = \frac{0.10 - 0.016}{2} = 0.042\ \text{m}$$

$$L_{k\,(12.5)} = \frac{T - L_U}{2} = \frac{0.125 - 0.016}{2} = 0.0545\ \text{m}$$

$$L_{k\,(15)} = \frac{T - L_U}{2} = \frac{0.15 - 0.016}{2} = 0.067\ \text{m}$$

$$L_{k\,(20)} = \frac{T - L_U}{2} = \frac{0.20 - 0.016}{2} = 0.092\ \text{m}$$

$$k_{k,xps\,(10)} = \frac{\ell}{L_k} \tanh\left(\frac{L_k}{\ell}\right) = \frac{0.0211}{0.042} \tanh\left(\frac{0.042}{0.0211}\right) = 0.484$$

$$k_{k,xps\,(12.5)} = \frac{\ell}{L_k} \tanh\left(\frac{L_k}{\ell}\right) = \frac{0.0211}{0.0545} \tanh\left(\frac{0.0545}{0.0211}\right) = 0.382$$

$$k_{k,xps\,(15)} = \frac{\ell}{L_k} \tanh\left(\frac{L_k}{\ell}\right) = \frac{0.0211}{0.067} \tanh\left(\frac{0.067}{0.0211}\right) = 0.313$$

$$k_{k,xps\,(20)} = \frac{\ell}{L_k} \tanh\left(\frac{L_k}{\ell}\right) = \frac{0.0211}{0.092} \tanh\left(\frac{0.092}{0.0211}\right) = 0.229$$

$$k_{k,eps\,(10)} = \frac{\ell}{L_k} \tanh\left(\frac{L_k}{\ell}\right) = \frac{0.0210}{0.042} \tanh\left(\frac{0.042}{0.0210}\right) = 0.481$$

$$k_{k,eps\,(12.5)} = \frac{\ell}{L_k} \tanh\left(\frac{L_k}{\ell}\right) = \frac{0.0210}{0.0545} \tanh\left(\frac{0.0545}{0.0210}\right) = 0.380$$

$$k_{k,eps\,(15)} = \frac{\ell}{L_k}\tanh\left(\frac{L_k}{\ell}\right) = \frac{0.0210}{0.067}\tanh\left(\frac{0.067}{0.0210}\right) = 0.312$$

$$k_{k,eps\,(20)} = \frac{\ell}{L_k}\tanh\left(\frac{L_k}{\ell}\right) = \frac{0.0210}{0.092}\tanh\left(\frac{0.092}{0.0210}\right) = 0.228$$

$$\ell_{B,xps} = \sqrt{\frac{s_{WB} \times \lambda_{WB}}{\frac{1}{R_i} + \frac{1}{R_e}}} = \sqrt{\frac{0.01 \times 0.2166}{\frac{1}{0.262} + \frac{1}{0.941}}} = 0.0211$$

$$\ell_{B,eps} = \sqrt{\frac{s_{WB} \times \lambda_{WB}}{\frac{1}{R_i} + \frac{1}{R_e}}} = \sqrt{\frac{0.01 \times 0.2166}{\frac{1}{0.262} + \frac{1}{0.898}}} = 0.0210$$

Where the conductive layer is a dry screed with the following technical parameters:

$s_{WB} = 0.01$ m – thickness of the conductive layer (dry screed)
$\lambda_{WB} = 0.2166$ W / mK – thermal conductivity values of the screed layer

The thermal resistance under the heat-conducting layer (dry screed) to the room below or outside is

$$R_{e,xps\,(10)} = R_{e,0} - 2 \times \Delta R_e = 1.05 - 2 \times 0.0753 = 0.902\ m^2K / W$$

$$R_{e,xps\,(12.5)} = R_{e,0} - 2 \times \Delta R_e = 1.05 - 2 \times 0.0602 = 0.932\ m^2K / W$$

$$R_{e,xps\,(15)} = R_{e,0} - 2 \times \Delta R_e = 1.05 - 2 \times 0.0502 = 0.952\ m^2K / W$$

$$R_{e,xps\,(20)} = R_{e,0} - 2 \times \Delta R_e = 1.05 - 2 \times 0.0376 = 0.977\ m^2K / W$$

$$R_{e,eps\,(10)} = R_{e,0} - 2 \times \Delta R_e = 1.00 - 2 \times 0.0711 = 0.861\ m^2K / W$$

$$R_{e,eps\,(12.5)} = R_{e,0} - 2 \times \Delta R_e = 1.00 - 2 \times 0.0569 = 0.890\ m^2K / W$$

$$R_{e,eps\,(15)} = R_{e,0} - 2 \times \Delta R_e = 1.00 - 2 \times 0.047 = 0.909\ m^2K / W$$

$$R_{e,eps\,(20)} = R_{e,0} - 2 \times \Delta R_e = 1.00 - 2 \times 0.0356 = 0.932\ m^2K / W$$

Small differences in the thermal resistance values of the layer transferring heat to the room below allowed for averaging. Therefore, the following values were adopted for further analysis:

$$R_{e,xps} = 0.941\ m^2K / W$$

$$R_{e,eps} = 0.898\ m^2K / W$$

$$\Delta R_{e,xps\,(10)} = \frac{d}{\lambda_{ins}} \times \frac{b}{T} = \frac{0.016}{0.034} \times \frac{0.016}{0.10} = 0.0753\ m^2K / W$$

$$\Delta R_{e,xps\,(12.5)} = \frac{d}{\lambda_{ins}} \times \frac{b}{T} = \frac{0.016}{0.034} \times \frac{0.016}{0.125} = 0.0602\ m^2K / W$$

$$\Delta R_{e,xps\,(15)} = \frac{d}{\lambda_{ins}} \times \frac{b}{T} = \frac{0.016}{0.034} \times \frac{0.016}{0.15} = 0.0502\ m^2K / W$$

$$\Delta R_{e,xps\,(20)} = \frac{d}{\lambda_{ins}} \times \frac{b}{T} = \frac{0.016}{0.034} \times \frac{0.016}{0.20} = 0.0376\ m^2K / W$$

$$\Delta R_{e,eps\,(10)} = \frac{d}{\lambda_{ins}} \times \frac{b}{T} = \frac{0.016}{0.036} \times \frac{0.016}{0.10} = 0.0711\ m^2K / W$$

$$\Delta R_{e,eps\,(12.5)} = \frac{d}{\lambda_{ins}} \times \frac{b}{T} = \frac{0.016}{0.036} \times \frac{0.016}{0.125} = 0.0569\ m^2K / W$$

$$\Delta R_{e,eps\,(15)} = \frac{d}{\lambda_{ins}} \times \frac{b}{T} = \frac{0.016}{0.036} \times \frac{0.016}{0.15} = 0.0474\ m^2K / W$$

$$\Delta R_{e,eps\,(20)} = \frac{d}{\lambda_{ins}} \times \frac{b}{T} = \frac{0.016}{0.036} \times \frac{0.016}{0.20} = 0.0356\ m^2K / W$$

$$R_{e,0,xps} = R_u + R_{se} = \frac{d_k}{\lambda_k} + R_{se} = \frac{0.03}{0.034} + 0.17 = 1.05\ m^2K / W$$

$$R_{e,0,eps} = R_u + R_{se} = \frac{d_k}{\lambda_k} + R_{se} = \frac{0.03}{0.036} + 0.17 = 1.00\ m^2K / W$$

The thermal resistance of the layer conducting heat to the heated room is

$$R_i = R_o + R_{con,i} + R_{si} = 0.019 + 0.15 + 0.093 = 0.262\ m^2K / W$$

The resistance for carefully shaped conductive layers permanently glued together with an insulation layer is $R_{con,i} = 0.10\ m^2K / W$. Since the thermal insulation without aluminium lamellas is not glued to the screed, a higher thermal resistance should be assumed for calculations, which is

$$R_{con,i} = 0.15\ m^2K/W$$

The thermal resistance of the floor layer (flooring cover with adhesive) above the conductive layer is

$$R_o = \frac{d_i}{\lambda_i} = \frac{0.002}{0.25} + \frac{0.005}{0.45} = 0.019\ m^2K/W$$

The heat transfer resistance on the floor surface was assumed in accordance with the standard [5], as in the case of most underfloor heating systems, which is

$$R_{si} = \frac{1}{\alpha_i} = 0.093\ m^2K/W$$

The temperature difference between the maximum permissible average temperature and the air temperature, when we substitute k_{CB} values depending on the heating pipe spacing, XPS or EPS thermal insulation type, and the room's impact zone, will be

- in the central zone

$$\bar{\theta}^{max}_{F,xps\,(10)} - \theta_{i,xps\,(10)} = k_{CB}\left(\theta^{max}_F - \theta_i\right) = 0.5661(29-20) = 5.094°C$$

$$\bar{\theta}^{max}_{F,xps\,(12.5)} - \theta_{i,xps\,(12.5)} = k_{CB}\left(\theta^{max}_F - \theta_i\right) = 0.4613(29-20) = 4.1521°C$$

$$\bar{\theta}^{max}_{F,xps\,(15)} - \theta_{i,xps\,(15)} = k_{CB}\left(\theta^{max}_F - \theta_i\right) = 0.3867(29-20) = 3.4799°C$$

$$\bar{\theta}^{max}_{F,xps\,(20)} - \theta_{i,xps\,(20)} = k_{CB}\left(\theta^{max}_F - \theta_i\right) = 0.2906(29-20) = 2.6158°C$$

$$\bar{\theta}^{max}_{F,eps\,(10)} - \theta_{i,eps\,(10)} = k_{CB}\left(\theta^{max}_F - \theta_i\right) = 0.5643(29-20) = 5.0785°C$$

$$\bar{\theta}^{max}_{F,eps\,(12.5)} - \theta_{i,eps\,(12.5)} = k_{CB}\left(\theta^{max}_F - \theta_i\right) = 0.4597(29-20) = 4.1375°C$$

$$\bar{\theta}^{max}_{F,eps\,(15)} - \theta_{i,eps\,(15)} = k_{CB}\left(\theta^{max}_F - \theta_i\right) = 0.3852(29-20) = 3.4671°C$$

$$\bar{\theta}^{max}_{F,eps\,(20)} - \theta_{i,eps\,(20)} = k_{CB}\left(\theta^{max}_F - \theta_i\right) = 0.2896(29-20) = 2.6061°C$$

- in the edge zone

$$\overline{\theta}_{F,xps\,(10)}^{max} - \theta_{i,xps\,(10)} = k_{CB}\left(\theta_F^{max} - \theta_i\right) = 0.566(35-20) = 8.4910°C$$

$$\overline{\theta}_{F,xps\,(12.5)}^{max} - \theta_{i,xps\,(12.5)} = k_{CB}\left(\theta_F^{max} - \theta_i\right) = 0.4613(35-20) = 6.9202°C$$

$$\overline{\theta}_{F,xps\,(15)}^{max} - \theta_{i,xps\,(15)} = k_{CB}\left(\theta_F^{max} - \theta_i\right) = 0.3867(35-20) = 5.7998°C$$

$$\overline{\theta}_{F,xps\,(20)}^{max} - \theta_{i,xps\,(20)} = k_{CB}\left(\theta_F^{max} - \theta_i\right) = 0.2906(35-20) = 4.3597°C$$

$$\overline{\theta}_{F,eps\,(10)}^{max} - \theta_{i,eps\,(10)} = k_{CB}\left(\theta_F^{max} - \theta_i\right) = 0.5643(35-20) = 8.4910°C$$

$$\overline{\theta}_{F,eps\,(12.5)}^{max} - \theta_{i,eps\,(12.5)} = k_{CB}\left(\theta_F^{max} - \theta_i\right) = 0.4597(35-20) = 6.9202°C$$

$$\overline{\theta}_{F,eps\,(15)}^{max} - \theta_{i,eps\,(15)} = k_{CB}\left(\theta_F^{max} - \theta_i\right) = 0.3852(35-20) = 5.7786°C$$

$$\overline{\theta}_{F,eps\,(20)}^{max} - \theta_{i,eps\,(20)} = k_{CB}\left(\theta_F^{max} - \theta_i\right) = 0.2896(35-20) = 4.3435°C$$

Ultimately, the calculated maximum heat flux density, depending on the coil spacing and the type of EPS or XPS thermal insulation, assuming an indoor temperature of 20°C and an impact zone of q_i^{max}, is

- in the central zone

$$q_{xps\,(10)}^{max} = 8.92\left(\overline{\theta}_F^{max} - \theta_i\right)^{1.1} = 8.92(5.0946)^{1.1} = 53.48\ W/m^2$$

$$q_{xps\,(12.5)}^{max} = 8.92\left(\overline{\theta}_F^{max} - \theta_i\right)^{1.1} = 8.92(4.1521)^{1.1} = 42.70\ W/m^2$$

$$q_{xps\,(15)}^{max} = 8.92\left(\overline{\theta}_F^{max} - \theta_i\right)^{1.1} = 8.92(3.4799)^{1.1} = 35.16\ W/m^2$$

$$q_{xps\,(20)}^{max} = 8.92\left(\overline{\theta}_F^{max} - \theta_i\right)^{1.1} = 8.92(2.6158)^{1.1} = 25.69\ W/m^2$$

$$q_{eps\,(10)}^{max} = 8.92\left(\overline{\theta}_F^{max} - \theta_i\right)^{1.1} = 8.92(5.0785)^{1.1} = 53.29\ W/m^2$$

$$q_{eps\,(12.5)}^{max} = 8.92\left(\overline{\theta}_F^{max} - \theta_i\right)^{1.1} = 8.92(4.1375)^{1.1} = 42.54\ W/m^2$$

$$q_{eps\,(15)}^{max} = 8.92\left(\overline{\theta}_F^{max} - \theta_i\right)^{1.1} = 8.92(3.4671)^{1.1} = 35.02\ W/m^2$$

$$q_{eps\,(20)}^{max} = 8.92\left(\overline{\theta}_F^{max} - \theta_i\right)^{1.1} = 8.92(2.6061)^{1.1} = 25.58\ W/m^2$$

- in the edge zone

$$q_{xps\,(10)}^{max} = 8.92\left(\overline{\theta}_F^{max} - \theta_i\right)^{1.1} = 8.92(8.4910)^{1.1} = 93.80\ W/m^2$$

$$q_{xps\,(12.5)}^{max} = 8.92\left(\overline{\theta}_F^{max} - \theta_i\right)^{1.1} = 8.92(6.9202)^{1.1} = 74.90\ W/m^2$$

$$q_{xps\,(15)}^{max} = 8.92\left(\overline{\theta}_F^{max} - \theta_i\right)^{1.1} = 8.92(5.7998)^{1.1} = 61.68\ W/m^2$$

$$q_{xps\,(20)}^{max} = 8.92\left(\overline{\theta}_F^{max} - \theta_i\right)^{1.1} = 8.92(4.3597)^{1.1} = 45.06\ W/m^2$$

$$q_{eps\,(10)}^{max} = 8.92\left(\overline{\theta}_F^{max} - \theta_i\right)^{1.1} = 8.92(8.4642)^{1.1} = 93.48\ W/m^2$$

$$q_{eps\,(12.5)}^{max} = 8.92\left(\overline{\theta}_F^{max} - \theta_i\right)^{1.1} = 8.92(6.8959)^{1.1} = 74.61\ W/m^2$$

$$q_{eps\,(15)}^{max} = 8.92\left(\overline{\theta}_F^{max} - \theta_i\right)^{1.1} = 8.92(5.7786)^{1.1} = 61.43\ W/m^2$$

$$q_{eps\,(20)}^{max} = 8.92\left(\overline{\theta}_F^{max} - \theta_i\right)^{1.1} = 8.92(4.3435)^{1.1} = 44.87\ W/m^2$$

By substituting the values calculated above, we calculate the balanced heat transfer coefficient K_H, depending on the thermal insulation used and the distance of the heating pipes,

$$K_{H,xps\,(10)} = \frac{q_i^{max}}{\theta_F^{max} - \theta_i} = \frac{53.43}{29 - 20} = 5.942\ W/m^2K$$

$$K_{H,xps\,(12.5)} = \frac{q_i^{max}}{\theta_F^{max} - \theta_i} = \frac{42.70}{29 - 20} = 4.1521\ W/m^2K$$

$$K_{H,xps\,(15)} = \frac{q_i^{max}}{\theta_F^{max} - \theta_i} = \frac{35.16}{29 - 20} = 3.4799\ W/m^2K$$

$$K_{H,xps\,(20)} = \frac{q_i^{max}}{\theta_F^{max} - \theta_i} = \frac{25.69}{29 - 20} = 2.6158\ W/m^2K$$

$$K_{H,eps\,(10)} = \frac{q_i^{max}}{\theta_F^{max} - \theta_i} = \frac{53.29}{29 - 20} = 5.0785\ W/m^2K$$

$$K_{H,eps\,(12.5)} = \frac{q_i^{max}}{\theta_F^{max} - \theta_i} = \frac{42.54}{29 - 20} = 4.1375\ W/m^2K$$

$$K_{H,eps\,(15)} = \frac{q_i^{max}}{\theta_F^{max} - \theta_i} = \frac{35.02}{29 - 20} = 3.4671\ W/m^2K$$

$$K_{H,eps\,(20)} = \frac{q_i^{max}}{\theta_F^{max} - \theta_i} = \frac{25.58}{29 - 20} = 2.6061\ W/m^2K$$

The final value of the calculated heat flux density q_{ob}, emitted from the second-layer layout of the heated floor with a dry screed and without aluminium lamellas, taking into account the assumed criteria of the adopted example variant of layers with a vinyl panel floor at different coil spacings, is

$$q_{ob} = K_H \times \Delta\theta_H\ [W/m^2]$$

$$\Delta\theta_H = \theta_H - \theta_i$$

a. with a coil spacing of 10 cm

$$q_{xps,10\,(28°)} = 5.942 \times (28 - 20) = 47.52\ W/m^2$$

$$q_{xps,10\,(35°)} = 5.942 \times (35 - 20) = 89.13\ W/m^2$$

$$q_{xps,10\ (40°)} = 5.942 \times (40 - 20) = 118.84\ W/m^2$$

$$q_{xps,10\ (45°)} = 5.942 \times (45 - 20) = 148.55\ W/m^2$$

$$q_{eps,10\ (28°)} = 5.922 \times (28 - 20) = 47.37\ W/m^2$$

$$q_{eps,10\ (35°)} = 5.922 \times (35 - 20) = 88.82\ W/m^2$$

$$q_{eps,10\ (40°)} = 5.922 \times (40 - 20) = 118.43\ W/m^2$$

$$q_{eps,10\ (45°)} = 5.922 \times (45 - 20) = 148.04\ W/m^2$$

b. with a coil spacing of 12.5 cm

$$q_{xps,12.5\ (28°)} = 4.745 \times (28 - 20) = 37.96\ W/m^2$$

$$q_{xps,12.5\ (35°)} = 4.745 \times (35 - 20) = 71.17\ W/m^2$$

$$q_{xps,12.5\ (40°)} = 4.745 \times (40 - 20) = 94.90\ W/m^2$$

$$q_{xps,12.5\ (45°)} = 4.745 \times (45 - 20) = 118.62\ W/m^2$$

$$q_{eps,12.5\ (28°)} = 4.726 \times (28 - 20) = 37.81\ W/m^2$$

$$q_{eps,12.5\ (35°)} = 4.726 \times (35 - 20) = 70.90\ W/m^2$$

$$q_{eps,12.5\ (40°)} = 4.726 \times (40 - 20) = 94.53\ W/m^2$$

$$q_{eps,12.5\ (45°)} = 4.726 \times (45 - 20) = 118.16\ W/m^2$$

c. with a coil spacing of 15 cm

$$q_{xps,15\ (28^\circ)} = 3.907 \times (28 - 20) = 31.26\ \text{W/m}^2$$

$$q_{xps,15\ (35^\circ)} = 3.907 \times (35 - 20) = 58.60\ \text{W/m}^2$$

$$q_{xps,15\ (40^\circ)} = 3.907 \times (40 - 20) = 78.14\ \text{W/m}^2$$

$$q_{xps,15\ (45^\circ)} = 3.907 \times (45 - 20) = 97.67\ \text{W/m}^2$$

$$q_{eps,15\ (28^\circ)} = 3.891 \times (28 - 20) = 31.13\ \text{W/m}^2$$

$$q_{eps,15\ (35^\circ)} = 3.891 \times (35 - 20) = 58.37\ \text{W/m}^2$$

$$q_{eps,15\ (40^\circ)} = 3.891 \times (40 - 20) = 77.83\ \text{W/m}^2$$

$$q_{eps,15\ (45^\circ)} = 3.891 \times (45 - 20) = 97.28\ \text{W/m}^2$$

d. with a coil spacing of 20 cm

$$q_{xps,20\ (28^\circ)} = 2.854 \times (28 - 20) = 22.83\ \text{W/m}^2$$

$$q_{xps,20\ (35^\circ)} = 2.854 \times (35 - 20) = 42.81\ \text{W/m}^2$$

$$q_{xps,20\ (40^\circ)} = 2.854 \times (40 - 20) = 57.09\ \text{W/m}^2$$

$$q_{xps,20\ (45^\circ)} = 2.854 \times (45 - 20) = 71.36\ \text{W/m}^2$$

$$q_{eps,20\ (28^\circ)} = 2.843 \times (28 - 20) = 22.74\ \text{W/m}^2$$

$$q_{eps,20\ (35^\circ)} = 2.843 \times (35 - 20) = 42.64\ \text{W/m}^2$$

$$q_{eps,20\,(40^\circ)} = 2.843 \times (40 - 20) = 56.8\ \mathrm{W/m^2}$$

$$q_{eps,20\,(45^\circ)} = 2.843 \times (45 - 20) = 71.06\ \mathrm{W/m^2}$$

The calculation example for one variant of the second heated floor layout with vinyl flooring and a BAUNORM FIBERCEMENT dry screed without aluminium lamellas provides the basis for calculating the remaining variants of the second multi-variant layer layout. The set of mathematical formulas adopted in the example variant was applied to the calculations of other variants of the heated floor with KNAUF BRIO and KNAUF DUAL FLOOR dry screeds, as well as other floorings. The following seven floorings were used for underfloor heating:

- PVC flooring with integrated XPO underlay
- three-layer wooden board + polyurethane adhesive
- laminate panels + smart underlay
- two-component microcement Microflex
- PVC flooring + Eurocol 640 Eurostar adhesive
- carpet flooring + Eurocol 540 Eurosafe Special adhesive
- flexible resin flooring

The results of the thermal power calculations for the second-layer layout of the heated floor with different dry screeds, without heat-dissipating aluminium lamellas, are summarised in Table 5.4.

TABLE 5.4
Design Values of Thermal Power for the Eight Flooring Covers of the Second Heated Floor Layout with Screeds, without Aluminium Lamellas [2]

Coil Spacing [cm]		Design Thermal Efficiency, q_{ob} [W/m²]											
		Knauf Brio				BAUNORM Fibercement				Knauf Dual Floor			
		Coil Supply Temperature [°C]											
		28	35	40	45	28	35	40	45	28	35	40	45
		PVC Floor Panels – Vinyl + Polyurethane Adhesive PU 560											
XPS	10	59.62	111.79	149.06	186.32	47.54	89.13	118.84	148.55	64.86	121.62	162.15	202.69
	12.5	49.55	92.91	123.88	154.85	37.96	71.17	94.90	118.62	55.22	103.54	138.05	172.57
	15	41.65	78.09	104.12	130.14	31.26	58.60	78.14	97.67	47.15	88.40	117.87	147.34
	20	30.87	57.88	77.17	96.47	22.83	42.81	57.09	71.36	35.47	66.51	88.68	110.85
EPS	10	59.45	111.47	148.63	185.79	47.37	88.82	118.43	148.04	64.70	121.32	161.76	202.20
	12.5	49.37	92.58	123.43	154.29	37.81	70.90	94.53	118.16	55.04	103.20	137.61	172.01
	15	41.48	77.78	103.70	129.63	31.13	58.37	77.83	97.28	46.97	88.07	117.42	146.78
	20	30.73	57.63	76.84	96.05	22.74	42.64	56.85	71.06	35.32	66.22	88.29	110.36

(Continued)

TABLE 5.4 (*Continued*)

Design Values of Thermal Power for the Eight Flooring Covers of the Second Heated Floor Layout with Screeds, without Aluminium Lamellas [2]

Coil Spacing [cm]		Design Thermal Efficiency, q_{ob} [W/m^2]											
		Knauf Brio				BAUNORM Fibercement				Knauf Dual Floor			
		Coil Supply Temperature [°C]											
		28	35	40	45	28	35	40	45	28	35	40	45
		PC Floor Panels – Vinyl with Integrated XPO Underlay											
XPS	10	60.08	112.64	150.19	187.74	47.98	89.96	119.95	149.94	65.28	122.41	163.21	204.01
	12.5	50.02	93.79	125.06	156.32	38.35	71.91	95.88	119.85	55.70	104.44	139.25	174.06
	15	42.09	78.92	105.23	131.54	31.59	59.24	78.99	98.73	47.63	89.30	119.07	148.83
	20	31.23	58.56	78.08	97.60	23.09	43.29	57.72	72.15	35.89	67.29	89.71	112.14
EPS	10	59.90	112.32	149.75	187.19	47.81	89.64	119.52	149.40	65.12	122.10	162.81	203.51
	12.5	49.84	93.45	124.60	155.75	38.20	71.63	95.50	119.38	55.52	104.09	138.79	173.48
	15	41.92	78.60	104.80	131.00	31.46	58.99	78.66	98.32	47.44	88.95	118.60	148.26
	20	31.09	58.30	77.73	97.16	22.99	43.10	57.47	71.84	35.73	66.99	89.31	111.64
		Three-Layer Wooden Board + Polyurethane Adhesive PU 560											
XPS	10	62.19	116.61	155.47	194.34	50.10	93.94	125.25	156.56	67.23	126.06	168.08	210.10
	12.5	52.27	98.01	130.68	163.35	40.27	75.51	100.67	125.84	57.94	108.64	144.86	181.07
	15	44.25	82.96	110.62	138.27	33.25	62.35	83.13	103.91	49.91	93.58	124.77	155.97
	20	33.01	61.89	82.52	103.14	24.33	45.62	60.83	76.03	37.90	71.07	94.76	118.45
EPS	10	62.00	116.24	154.99	193.74	49.90	93.57	124.75	155.94	67.06	125.73	167.64	209.55
	12.5	52.06	97.	130.16	162.70	40.09	75.17	100.22	125.28	57.74	108.26	144.34	180.43
	15	44.04	82.58	110.11	137.64	33.09	62.05	82.74	103.42	49.70	93.18	124.24	155.30
	20	32.84	61.57	82.09	102.62	24.21	45.40	60.53	75.66	37.71	70.71	94.28	117.85
		Laminate Flooring + Smart Underlay											
XPS	10	64.40	120.75	161.00	201.25	52.42	98.29	131.05	163.81	69.23	129.81	173.07	216.34
	12.5	54.70	102.57	136.76	170.95	42.42	79.55	106.06	132.58	60.32	113.10	150.80	188.50
	15	46.63	87.43	116.58	145.72	35.14	65.89	87.86	109.82	52.39	98.23	130.97	163.72
	20	35.03	65.67	87.56	109.45	25.77	48.31	64.41	80.52	40.17	75.33	100.43	125.54
EPS	10	64.18	120.35	160.46	200.58	52.19	97.85	130.47	163.09	69.04	129.44	172.59	215.74
	12.5	54.46	102.12	136.16	170.19	42.21	79.14	105.52	131.90	60.09	112.66	150.22	187.77
	15	46.39	86.99	115.98	144.98	34.95	65.53	87.38	109.22	52.14	97.77	130.36	162.95
	20	34.82	65.29	87.05	108.81	25.62	48.03	64.05	80.06	39.94	74.90	99.86	124.83
		Microflex Two-Component Microcement											
XPS	10	58.83	110.31	147.08	183.86	46.78	87.70	116.94	146.17	64.12	120.23	160.30	200.38
	12.5	48.73	91.38	121.84	152.29	37.28	69.91	93.21	116.51	54.39	101.99	135.98	169.98
	15	40.88	76.65	102.20	127.75	30.68	57.52	76.70	95.87	46.32	86.86	115.81	144.76
	20	30.25	56.72	75.63	94.54	22.40	42.01	56.01	70.01	34.76	65.18	86.90	108.63
EPS	10	58.67	110.01	146.68	183.35	46.62	87.41	116.55	145.69	63.97	119.94	159.92	199.90
	12.5	48.57	91.06	121.42	151.77	37.15	69.65	92.87	116.08	54.22	101.67	135.55	169.44
	15	40.72	76.36	101.81	127.26	30.56	57.30	76.40	95.51	46.15	86.54	115.38	144.23
	20	30.13	56.49	75.31	94.14	22.32	41.85	55.79	69.74	34.62	64.91	86.54	108.18

(*Continued*)

TABLE 5.4 (*Continued*)
Design Values of Thermal Power for the Eight Flooring Covers of the Second Heated Floor Layout with Screeds, without Aluminium Lamellas [2]

Coil Spacing [cm]		Design Thermal Efficiency, q_{ob} [W/m^2]											
		Knauf Brio				BAUNORM Fibercement				Knauf Dual Floor			
		Coil Supply Temperature [°C]											
		28	35	40	45	28	35	40	45	28	35	40	45
		PVC Flooring + Eurocol 640 Eurostar Special Adhesive											
XPS	10	59.47	111.51	148.67	185.84	47.39	88.85	118.47	148.09	64.72	121.35	161.80	202.24
	12.5	49.39	92.61	123.48	154.35	37.83	70.92	94.56	118.21	55.06	103.24	137.65	172.06
	15	41.50	77.80	103.74	129.67	31.14	58.39	77.86	97.32	46.99	88.10	117.47	146.83
	20	30.75	57.65	76.87	96.09	22.75	42.66	56.87	71.09	35.33	66.25	88.33	110.41
EPS	10	59.26	111.12	148.15	185.19	47.22	88.55	118.06	147.58	64.51	120.96	161.28	201.59
	12.5	49.22	92.28	123.04	153.80	37.68	70.65	94.20	117.75	54.88	102.90	137.21	171.51
	15	41.33	77.50	103.33	129.16	31.02	58.16	77.55	96.93	46.81	87.77	117.02	146.28
	20	30.61	57.40	76.54	95.67	22.66	42.48	56.64	70.80	35.18	65.96	87.95	109.93
		Flotex Colour Carpet + Eurocol 640 Eurosafe Special Adhesive											
XPS	10	62.20	116.62	155.50	194.37	50.11	93.95	125.27	156.59	67.24	126.08	168.10	210.13
	12.5	52.28	98.03	130.70	163.38	40.28	75.52	100.69	125.87	57.95	108.66	144.88	181.10
	15	44.26	82.98	110.64	138.30	33.26	62.36	83.15	103.94	49.92	93.60	124.80	156.00
	20	33.01	61.90	82.54	103.17	24.34	45.63	60.84	76.05	37.91	71.09	94.78	118.48
EPS	10	62.00	116.26	155.01	193.76	49.91	93.58	124.78	155.97	67.06	125.74	167.66	209.57
	12.5	52.07	97.64	130.18	162.73	40.10	75.18	100.24	125.30	57.75	108.27	144.37	180.46
	15	44.05	82.60	110.13	137.67	33.10	62.07	82.75	103.44	49.71	93.20	124.26	155.33
	20	32.85	61.58	82.11	102.64	24.22	45.41	60.54	75.68	37.72	70.73	94.31	117.88
		Sikafloor Elastic Resin Floor – 300 N + Sikafloor 156 Primer											
XPS	10	59.43	111.43	148.57	185.71	47.35	88.77	118.37	147.96	64.68	121.27	161.70	202.12
	12.5	49.35	92.52	123.37	154.21	37.79	70.85	94.47	118.09	55.01	103.15	137.54	171.92
	15	41.45	77.73	103.64	129.54	31.11	58.33	77.78	97.22	46.94	88.02	117.35	146.69
	20	30.71	57.59	76.78	95.98	22.73	42.61	56.81	71.02	35.29	66.17	88.23	110.29
EPS	10	59.26	111.11	148.15	185.18	47.18	88.47	117.96	147.45	64.52	120.98	161.30	201.63
	12.5	49.17	92.20	122.93	153.66	37.64	70.58	94.11	117.64	54.84	102.82	137.09	171.37
	15	41.29	77.42	103.23	129.03	30.99	58.10	77.47	96.84	46.77	87.68	116.91	146.14
	20	30.58	57.34	76.45	95.57	22.63	42.44	56.59	70.73	35.14	65.89	87.85	109.81

5.1.4 Determination of the Design Thermal Power of a *Lightweight* Underfloor Heating Layout *without Screed and without Lamellas*

The third heated floor layout is a typical lightweight heated floor system, with no screed and no metal lamellas. This system was subjected to mathematical calculations similar to the second layout, using the formulas adopted in [34]. The example calculations assumed the same floor made of glued vinyl panels as in the first and second screed layouts. This cover flooring was bonded to a self-levelling screed layer using a dedicated polyurethane adhesive supplied by the vinyl panel manufacturer, over a supporting composite adhesive-mesh layer. The remaining material assumptions, technical guidelines, and heat output calculation variants were unchanged and were based on the formulas developed in [34]. This will allow a comparison of the final calculated thermal output values for a multi-variant selection of different flooring and dry screed types, with and without heat-dissipating metal lamellas, and with screeds eliminated. In the third layout, similarly to the second layout of heated floors without lamellas, the maximum water temperature in the coil was not determined. Currently, there is no procedure for reliably calculating this maximum temperature in a lightweight heated floor system, and the calculation procedure based on the Scandinavian standard [5] is not applicable. As in the second layout, all components of the formulas given below were already listed and described in Chapter 4 on the calculation of thermal output of floors without lamellas; therefore, a detailed description of their meaning has not been undertaken again. An example of heat flux density calculations for a selected variant of a heated floor without screeds was taken from a recent engineering thesis [2], written under the supervision of the author of this monograph. Importantly, in the third configuration, the thermally conductive layer was an adhesive-mesh carrier, which also ensured the stiffness and stability of the floor structure. As in the previous considerations, in the example calculations, glued vinyl panels were assumed as the finishing layer, and the layer system was also devoid of heat-conducting lamellas. A significant difference in the calculations for the third-layer layout was that the heat transfer resistance between the adhesive-mesh layer and the thermal insulation was reduced, as these two layers were bonded together and there were no air gaps.

1. The calculation procedure for the theoretical maximum heat flux density of the third layout of a lightweight heated floor without lamellas, depending on the distance of the heating pipes, with detailed technical data of the individual layers, is presented below:
   - PVC vinyl floor panels, 5 mm thick, $\lambda = 0.45$ W/(m × K) [13],
   - PU 560 polyurethane adhesive, 2 mm thick, $\lambda = 0.25$ W/(m × K), [14],
   - mineral self-levelling layer, 5 mm thick, $\lambda = 0.85$ W/(m × K) [35],
   - adhesive-mesh layer, 3 mm thick, $\lambda = 0.85$ W/(m × K) [31],
   - multilayer heating pipe with aluminium insert, d = 16 mm, s = 2 mm, $\lambda = 0.43$ W/(m × K) [33],
   - Waterpanel X XPS insulation board, thick 30 mm, $\lambda = 0.034$ W/(m × K) or EPS Waterpanel E, thickness 30 mm, $\lambda = 0.036$ W/(m × K) [31].

$$q_B^{max} = 8.92\left(\overline{\theta}_F^{max} - \theta_i\right)^{1.1} \ [\mathrm{W/m^2}]$$

$$\overline{\theta}_F^{max} - \theta_i = k_{CB}\left(\theta_F^{max} - \theta_i\right)$$

The thermal conductivity coefficient of the dissipating layer (lamella) k_{CB} for different coil spacings is

$$k_{CB,xps\,(10)} = \frac{L_U + 2L_k \times k_k}{T} = \frac{0.016 + 2 \times 0.042 \times 0.687}{0.10} = 0.7370\ W/m^2K$$

$$k_{CB,xps\,(12.5)} = \frac{L_U + 2L_k \times k_k}{T} = \frac{0.016 + 2 \times 0.0545 \times 0.579}{0.125} = 0.6330\ W/m^2K$$

$$k_{CB,xps\,(15)} = \frac{L_U + 2L_k \times k_k}{T} = \frac{0.016 + 2 \times 0.067 \times 0.492}{0.15} = 0.5460\ W/m^2K$$

$$k_{CB,xps\,(20)} = \frac{L_U + 2L_k \times k_k}{T} = \frac{0.016 + 2 \times 0.092 \times 0.369}{0.20} = 0.4199\ W/m^2K$$

$$k_{CB,\,eps\,(10)} = \frac{L_U + 2L_k \times k_k}{T} = \frac{0.016 + 2 \times 0.042 \times 0.685}{0.10} = 0.7355\ W/m^2K$$

$$k_{CB,\,eps\,(12.5)} = \frac{L_U + 2L_k \times k_k}{T} = \frac{0.016 + 2 \times 0.0545 \times 0.577}{0.125} = 0.6313\ W/m^2K$$

$$k_{CB,\,eps\,(15)} = \frac{L_U + 2L_k \times k_k}{T} = \frac{0.016 + 2 \times 0.067 \times 0.490}{0.15} = 0.5444\ W/m^2K$$

$$k_{CB,\,eps\,(20)} = \frac{L_U + 2L_k \times k_k}{T} = \frac{0.016 + 2 \times 0.092 \times 0.368}{0.20} = 0.4185\ W/m^2K$$

An integral element of the formula for the thermal conductivity coefficient of the adhesive layer and the cover flooring without lamellas k_{CB} are the calculation below for different coil spacings and assuming no gap between the adhesive and the flooring:

$$L_{k\,(10)} = \frac{T - L_U}{2} = \frac{0.10 - 0.016}{2} = 0.042\ m$$

$$L_{k\,(12.5)} = \frac{T - L_U}{2} = \frac{0.125 - 0.016}{2} = 0.0545\ m$$

$$L_{k\,(15)} = \frac{T - L_U}{2} = \frac{0.15 - 0.016}{2} = 0.067\ m$$

$$L_{k\,(20)} = \frac{T - L_U}{2} = \frac{0.20 - 0.016}{2} = 0.092\ m$$

$$k_{k,xps\,(10)} = \frac{\ell}{L_k}\tanh\left(\frac{L_k}{\ell}\right) = \frac{0.0343}{0.042}\tanh\left(\frac{0.042}{0.0343}\right) = 0.687$$

$$k_{k,xps\,(12.5)} = \frac{\ell}{L_k}\tanh\left(\frac{L_k}{\ell}\right) = \frac{0.0343}{0.0545}\tanh\left(\frac{0.0545}{0.0343}\right) = 0.579$$

$$k_{k,xps\,(15)} = \frac{\ell}{L_k}\tanh\left(\frac{L_k}{\ell}\right) = \frac{0.0343}{0.067}\tanh\left(\frac{0.067}{0.0343}\right) = 0.492$$

$$k_{k,xps\,(20)} = \frac{\ell}{L_k}\tanh\left(\frac{L_k}{\ell}\right) = \frac{0.0343}{0.092}\tanh\left(\frac{0.092}{0.0343}\right) = 0.369$$

$$k_{k,eps\,(10)} = \frac{\ell}{L_k}\tanh\left(\frac{L_k}{\ell}\right) = \frac{0.0342}{0.042}\tanh\left(\frac{0.042}{0.0342}\right) = 0.685$$

$$k_{k,eps\,(12.5)} = \frac{\ell}{L_k}\tanh\left(\frac{L_k}{\ell}\right) = \frac{0.0342}{0.0545}\tanh\left(\frac{0.0545}{0.0342}\right) = 0.577$$

$$k_{k,eps\,(15)} = \frac{\ell}{L_k}\tanh\left(\frac{L_k}{\ell}\right) = \frac{0.0342}{0.067}\tanh\left(\frac{0.067}{0.0342}\right) = 0.490$$

$$k_{k,eps\,(20)} = \frac{\ell}{L_k}\tanh\left(\frac{L_k}{\ell}\right) = \frac{0.0342}{0.092}\tanh\left(\frac{0.092}{0.0342}\right) = 0.368$$

$$\ell_{B,xps} = \sqrt{\frac{s_{WB} \times \lambda_{WB}}{\frac{1}{R_i} + \frac{1}{R_e}}} = \sqrt{\frac{0.008 \times 0.85}{\frac{1}{0.212} + \frac{1}{0.941}}} = 0.0343$$

$$\ell_{B,eps} = \sqrt{\frac{s_{WB} \times \lambda_{WB}}{\frac{1}{R_i} + \frac{1}{R_e}}} = \sqrt{\frac{0.008 \times 0.85}{\frac{1}{0.212} + \frac{1}{0.898}}} = 0.0342$$

where

$s_{WB} = 0.08$ m – thickness of the conductive layer (adhesive + self-levelling screed)

$\lambda_{WB} = 1.0$ W / mK – weight average of the thermal conductivity values of the adhesive layer and self-levelling screed, taking into account their thickness

$$\lambda_{WB} = \frac{3\text{ mm}}{8\text{ mm}} \times 0.85 + \frac{5\text{ mm}}{8\text{ mm}} \times 0.85 = 0.21 + 0.79 = 1.0$$

The thermal resistance under the heat-conducting layer to the room below or outside is

$$R_{e,xps\,(10)} = R_{e,0} - 2\times \Delta R_e = 1.05 - 2\times 0.0753 = 0.902\ m^2K/W$$

$$R_{e,xps\,(12.5)} = R_{e,0} - 2\times \Delta R_e = 1.05 - 2\times 0.0602 = 0.932\ m^2K/W$$

$$R_{e,xps\,(15)} = R_{e,0} - 2\times \Delta R_e = 1.05 - 2\times 0.0502 = 0.952\ m^2K/W$$

$$R_{e,xps\,(20)} = R_{e,0} - 2\times \Delta R_e = 105 - 2\times 00376 = 0.977\ m^2K/W$$

$$R_{e,eps\,(10)} = R_{e,0} - 2\times \Delta R_e = 100 - 2\times 0.0711 = 0.861\ m^2K/W$$

$$R_{e,eps\,(12.5)} = R_{e,0} - 2\times \Delta R_e = 1.00 - 2\times 0.0569 = 0.890\ m^2K/W$$

$$R_{e,eps\,(15)} = R_{e,0} - 2\times \Delta R_e = 1.00 - 2\times 0.047 = 0.909\ m^2K/W$$

$$R_{e,eps\,(20)} = R_{e,0} - 2\times \Delta R_e = 1.00 - 2\times 0.0356 = 0.932\ m^2K/W$$

Small differences in the thermal resistance values of the layer transferring heat to the room below allowed for averaging. Therefore, the following values were adopted for further analysis:

$$R_{e,xps} = 0.941\ m^2K/W$$

$$R_{e,eps} = 0.898\ m^2K/W$$

$$\Delta R_{e,xps\,(10)} = \frac{d}{\lambda_{ins}} \times \frac{b}{T} = \frac{0.016}{0.034} \times \frac{0.016}{0.10} = 0.0753\ m^2K/W$$

$$\Delta R_{e,xps\,(12.5)} = \frac{d}{\lambda_{ins}} \times \frac{b}{T} = \frac{0.016}{0.034} \times \frac{0.016}{0.125} = 0.0602\ m^2K/W$$

$$\Delta R_{e,xps\,(15)} = \frac{d}{\lambda_{ins}} \times \frac{b}{T} = \frac{0.016}{0.034} \times \frac{0.016}{0.15} = 0.0502\ m^2K/W$$

$$\Delta R_{e,xps\,(20)} = \frac{d}{\lambda_{ins}} \times \frac{b}{T} = \frac{0.016}{0.034} \times \frac{0.016}{0.20} = 0.0376\ m^2K/W$$

$$\Delta R_{e,eps\,(10)} = \frac{d}{\lambda_{ins}} \times \frac{b}{T} = \frac{0.016}{0.036} \times \frac{0.016}{0.10} = 0.0711\ m^2K/W$$

$$\Delta R_{e,eps\,(12.5)} = \frac{d}{\lambda_{ins}} \times \frac{b}{T} = \frac{0.016}{0.036} \times \frac{0.016}{0.125} = 0.0569\ m^2K/W$$

$$\Delta R_{e,eps\,(15)} = \frac{d}{\lambda_{ins}} \times \frac{b}{T} = \frac{0.016}{0.036} \times \frac{0.016}{0.15} = 0.0474\ m^2K/W$$

$$\Delta R_{e,eps\,(20)} = \frac{d}{\lambda_{ins}} \times \frac{b}{T} = \frac{0.016}{0.036} \times \frac{0.016}{0.20} = 0.0356\ m^2K/W$$

$$R_{e,0,xps} = R_u + R_{se} = \frac{d_k}{\lambda_k} + R_{se} = \frac{0.03}{0.034} + 0.17 = 1.05\ m^2K/W$$

$$R_{e,0,eps} = R_u + R_{se} = \frac{d_k}{\lambda_k} + R_{se} = \frac{0.03}{0.036} + 0.17 = 1.00\ m^2K/W$$

The thermal resistance of the layer conducting heat to the heated room is

$$R_i = R_o + R_{con,i} + R_{si} = 0.019 + 0.10 + 0.093 = 0.212\ m^2K/W$$

The resistance for carefully shaped conductive layers permanently bonded to an insulation layer is $R_{con,i} = 0.10\ m^2K/W$. Because the thermal insulation is bonded to the supporting adhesive-mesh layer, this thermal resistance value of $R_{con,i} = 0.10\ m^2K/W$ should be used for calculations, unlike the values previously assumed in the first and second layout.

The thermal resistance of the cover flooring layers (floor with adhesive) above the conductive layer is

$$R_o = \frac{d_i}{\lambda_i} = \frac{0.002}{0.25} + \frac{0.005}{0.45} = 0.019\ m^2K/W$$

The heat transfer resistance on the floor surface was assumed in accordance with the standard [5], as in the case of most underfloor heating systems, which is

$$R_{si} = \frac{1}{\alpha_i} = 0.093\ m^2K/W$$

The temperature difference between the maximum permissible average temperature and the air temperature, when we substitute the k_{CB} values depending on the spacing of heating pipes, the type of XPS or EPS thermal insulation and the room impact zone, will be

- in the central zone

$$\overline{\theta}_{F,xps\,(10)}^{max} - \theta_{i,xps\,(10)} = k_{CB}\left(\theta_F^{max} - \theta_i\right) = 0.7370(29-20) = 6.6328°C$$

$$\overline{\theta}_{F,xps\,(12.5)}^{max} - \theta_{i,xps\,(12.5)} = k_{CB}\left(\theta_F^{max} - \theta_i\right) = 0.6330(29-20) = 5.6967°C$$

$$\overline{\theta}_{F,xps\,(15)}^{max} - \theta_{i,xps\,(15)} = k_{CB}\left(\theta_F^{max} - \theta_i\right) = 0.5460(29-20) = 4.9144°C$$

$$\overline{\theta}_{F,xps\,(20)}^{max} - \theta_{i,xps\,(20)} = k_{CB}\left(\theta_F^{max} - \theta_i\right) = 0.4199(29-20) = 3.7788°C$$

$$\overline{\theta}_{F,eps\,(10)}^{max} - \theta_{i,eps\,(10)} = k_{CB}\left(\theta_F^{max} - \theta_i\right) = 0.7355(29-20) = 6.6197°C$$

$$\overline{\theta}_{F,eps\,(12.5)}^{max} - \theta_{i,eps\,(12.5)} = k_{CB}\left(\theta_F^{max} - \theta_i\right) = 0.6313(29-20) = 5.6821°C$$

$$\overline{\theta}_{F,eps\,(15)}^{max} - \theta_{i,eps\,(15)} = k_{CB}\left(\theta_F^{max} - \theta_i\right) = 0.5444(29-20) = 4.8999°C$$

$$\overline{\theta}_{F,eps\,(20)}^{max} - \theta_{i,eps\,(20)} = k_{CB}\left(\theta_F^{max} - \theta_i\right) = 0.4185(29-20) = 3.7662°C$$

- in the edge zone

$$\overline{\theta}_{F,xps\,(10)}^{max} - \theta_{i,xps\,(10)} = k_{CB}\left(\theta_F^{max} - \theta_i\right) = 0.7370(35-20) = 11.0546°C$$

$$\overline{\theta}_{F,xps\,(12.5)}^{max} - \theta_{i,xps\,(12.5)} = k_{CB}\left(\theta_F^{max} - \theta_i\right) = 0.6330(35-20) = 9.4944°C$$

$$\overline{\theta}_{F,xps\,(15)}^{max} - \theta_{i,xps\,(15)} = k_{CB}\left(\theta_F^{max} - \theta_i\right) = 0.5460(35-20) = 8.1907°C$$

$$\overline{\theta}_{F,xps\,(20)}^{max} - \theta_{i,xps\,(20)} = k_{CB}\left(\theta_F^{max} - \theta_i\right) = 0.4199(35-20) = 6.2980°C$$

$$\overline{\theta}_{F,eps\,(10)}^{max} - \theta_{i,eps\,(10)} = k_{CB}\left(\theta_F^{max} - \theta_i\right) = 0.7355(35-20) = 11.0329°C$$

$$\overline{\theta}_{F,eps\,(12.5)}^{max} - \theta_{i,eps\,(12.5)} = k_{CB}\left(\theta_F^{max} - \theta_i\right) = 0.6313(35-20) = 9.4701°C$$

$$\overline{\theta}^{max}_{F,eps\,(15)} - \theta_{i,eps\,(15)} = k_{CB}\left(\theta^{max}_{F} - \theta_i\right) = 0.5444(35-20) = 8.1664°C$$

$$\overline{\theta}^{max}_{F,eps\,(20)} - \theta_{i,eps\,(20)} = k_{CB}\left(\theta^{max}_{F} - \theta_i\right) = 0.4185(35-20) = 6.2769°C$$

Finally, the calculated maximum heat flux density depending on the coil spacing, the type of thermal insulation, EPS or XPS, assuming the temperature inside the room is 20°C, and the impact zone in the room q^{max}_{i} is

- in the central zone

$$q^{max}_{xps\,(10)} = 8.92\left(\overline{\theta}^{max}_{F} - \theta_i\right)^{1.1} = 8.92(6.6328)^{1.1} = 71.49\ W/m^2$$

$$q^{max}_{xps\,(12.5)} = 8.92\left(\overline{\theta}^{max}_{F} - \theta_i\right)^{1.1} = 8.92(5.6967)^{1.1} = 60.47\ W/m^2$$

$$q^{max}_{xps\,(15)} = 8.92\left(\overline{\theta}^{max}_{F} - \theta_i\right)^{1.1} = 8.92(4.9144)^{1.1} = 51.40\ W/m^2$$

$$q^{max}_{xps\,(20)} = 8.92\left(\overline{\theta}^{max}_{F} - \theta_i\right)^{1.1} = 8.92(3.7788)^{1.1} = 38.50\ W/m^2$$

$$q^{max}_{eps\,(10)} = 8.92\left(\overline{\theta}^{max}_{F} - \theta_i\right)^{1.1} = 8.92(6.6197)^{1.1} = 71.33\ W/m^2$$

$$q^{max}_{eps\,(12.5)} = 8.92\left(\overline{\theta}^{max}_{F} - \theta_i\right)^{1.1} = 8.92(5.6821)^{1.1} = 60.30\ W/m^2$$

$$q^{max}_{eps\,(15)} = 8.92\left(\overline{\theta}^{max}_{F} - \theta_i\right)^{1.1} = 8.92(4.8999)^{1.1} = 51.24\ W/m^2$$

$$q^{max}_{eps\,(20)} = 8.92\left(\overline{\theta}^{max}_{F} - \theta_i\right)^{1.1} = 8.92(3.7662)^{1.1} = 38.36\ W/m^2$$

- in the edge zone

$$q^{max}_{xps\,(10)} = 8.92\left(\overline{\theta}^{max}_{F} - \theta_i\right)^{1.1} = 8.92(11.0546)^{1.1} = 125.39\ W/m^2$$

$$q^{max}_{xps\,(12.5)} = 8.92\left(\overline{\theta}^{max}_{F} - \theta_i\right)^{1.1} = 8.92(9.4944)^{1.1} = 106.07\ W/m^2$$

$$q_{xps\,(15)}^{max} = 8.92\left(\overline{\theta}_F^{max} - \theta_i\right)^{1.1} = 8.92(8.1907)^{1.1} = 90.16\ W/\ m^2$$

$$q_{xps\,(20)}^{max} = 8.92\left(\overline{\theta}_F^{max} - \theta_i\right)^{1.1} = 8.92(6.2980)^{1.1} = 67.53\ W/\ m^2$$

$$q_{eps\,(10)}^{max} = 8.92\left(\overline{\theta}_F^{max} - \theta_i\right)^{1.1} = 8.92(11.0546)^{1.1} = 125.12\ W/\ m^2$$

$$q_{eps\,(12.5)}^{max} = 8.92\left(\overline{\theta}_F^{max} - \theta_i\right)^{1.1} = 8.92(9.4944)^{1.1} = 105.77\ W/\ m^2$$

$$q_{eps\,(15)}^{max} = 8.92\left(\overline{\theta}_F^{max} - \theta_i\right)^{1.1} = 8.92(8.1907)^{1.1} = 89.87\ W/\ m^2$$

$$q_{eps\,(20)}^{max} = 8.92\left(\overline{\theta}_F^{max} - \theta_i\right)^{1.1} = 8.92(6.2769)^{1.1} = 67.28\ W/\ m^2$$

By substituting the values calculated above, we calculate the balanced heat transfer coefficient K_H, depending on the thermal insulation used and the distance of the heating pipes,

$$K_{H,xps\,(10)} = \frac{q_i^{max}}{\theta_F^{max} - \theta_i} = \frac{71.49}{29 - 20} = 7.943\ W/m^2K$$

$$K_{H,xps\,(12.5)} = \frac{q_i^{max}}{\theta_F^{max} - \theta_i} = \frac{60.47}{29 - 20} = 6.719\ W/m^2K$$

$$K_{H,xps\,(15)} = \frac{q_i^{max}}{\theta_F^{max} - \theta_i} = \frac{51.40}{29 - 20} = 5.711\ W/m^2K$$

$$K_{H,xps\,(20)} = \frac{q_i^{max}}{\theta_F^{max} - \theta_i} = \frac{38.50}{29 - 20} = 4.278\ W/m^2K$$

$$K_{H,eps\,(10)} = \frac{q_i^{max}}{\theta_F^{max} - \theta_i} = \frac{71.33}{29 - 20} = 7.926\ W/m^2K$$

$$K_{H,eps\,(12.5)} = \frac{q_i^{max}}{\theta_F^{max} - \theta_i} = \frac{60.30}{29 - 20} = 6.700\ W/m^2K$$

$$K_{H,eps\,(15)} = \frac{q_i^{max}}{\theta_F^{max} - \theta_i} = \frac{51.24}{29 - 20} = 5.693\ W/m^2K$$

$$K_{H,eps\,(20)} = \frac{q_i^{max}}{\theta_F^{max} - \theta_i} = \frac{38.36}{29 - 20} = 4.262\ W/m^2K$$

The final value of the calculated heat flux density q_{obw}, emitted from the third-layer layout of the heated floor without screed and without aluminium lamellas, taking into account the assumed criteria of the adopted example variant of layers with a vinyl panel floor and a supporting adhesive-mesh layer, with different coil spacings, is

$$q_{obw} = K_H \times \Delta\theta_H\ [W/m^2]$$

$$\Delta\theta_H = \theta_H - \theta_i$$

a. with a coil spacing of 10 cm

$$q_{xps,10\,(28°)} = 7.943 \times (28 - 20) = 63.54\ W/m^2$$

$$q_{xps,10\,(35°)} = 7.943 \times (35 - 20) = 119.15\ W/m^2$$

$$q_{xps,10\,(40°)} = 7.943 \times (40 - 20) = 158.86\ W/m^2$$

$$q_{xps,10\,(45°)} = 7.943 \times (45 - 20) = 198.58\ W/m^2$$

$$q_{eps,10\,(28°)} = 7.926 \times (28 - 20) = 63.41\ W/m^2$$

$$q_{eps,10\,(35°)} = 7.926 \times (35 - 20) = 118.89\ W/m^2$$

$$q_{eps,10\,(40°)} = 7.926 \times (40 - 20) = 158.52\ W/m^2$$

$$q_{eps,10\,(45°)} = 7.926 \times (45 - 20) = 198.15\ W/m^2$$

b. with a coil spacing of 12.5 cm

$$q_{xps,12.5\,(28°)} = 6.719 \times (28 - 20) = 53.75\ W/m^2$$

$$q_{xps,12.5\,(35°)} = 6.719 \times (35 - 20) = 100.78\ W/m^2$$

$$q_{xps,12.5\ (40°)} = 6.719 \times (40 - 20) = 134.38\ \text{W/m}^2$$

$$q_{xps,12.5\ (45°)} = 6.719 \times (45 - 20) = 167.97\ \text{W/m}^2$$

$$q_{eps,12.5\ (28°)} = 6.700 \times (28 - 20) = 53.60\ \text{W/m}^2$$

$$q_{eps,12.5\ (35°)} = 6.700 \times (35 - 20) = 100.50\ \text{W/m}^2$$

$$q_{eps,12.5\ (40°)} = 6.700 \times (40 - 20) = 134.00\ \text{W/m}^2$$

$$q_{eps,12.5\ (45°)} = 6.700 \times (45 - 20) = 167.50\ \text{W/m}^2$$

c. with a coil spacing of 15 cm

$$q_{xps,15\ (28°)} = 5.711 \times (28 - 20) = 45.69\ \text{W/m}^2$$

$$q_{xps,15\ (35°)} = 5.711 \times (35 - 20) = 85.67\ \text{W/m}^2$$

$$q_{xps,15\ (40°)} = 5.711 \times (40 - 20) = 114.23\ \text{W/m}^2$$

$$q_{xps,15\ (45°)} = 5.711 \times (45 - 20) = 142.78\ \text{W/m}^2$$

$$q_{eps,15\ (28°)} = 5.693 \times (28 - 20) = 45.54\ \text{W/m}^2$$

$$q_{eps,15\ (35°)} = 5.693 \times (35 - 20) = 85.39\ \text{W/m}^2$$

$$q_{eps,15\ (40°)} = 5.693 \times (40 - 20) = 113.86\ \text{W/m}^2$$

$$q_{eps,15\ (45°)} = 5.693 \times (45 - 20) = 142.32\ \text{W/m}^2$$

d. with a coil spacing of 20 cm

$$q_{xps,20\ (28°)} = 4.278 \times (28 - 20) = 34.22\ \text{W/m}^2$$

$$q_{xps,20\ (35°)} = 4.278 \times (35 - 20) = 64.17\ \text{W/m}^2$$

$$q_{xps,20\ (40°)} = 4.278 \times (40 - 20) = 85.55\ \text{W/m}^2$$

$$q_{xps,20(45°)} = 4.278 \times (45 - 20) = 106.94\ \text{W/m}^2$$

$$q_{eps,20\ (28°)} = 4.262 \times (28 - 20) = 34.10\ \text{W/m}^2$$

$$q_{eps,20\ (35°)} = 4.262 \times (35 - 20) = 63.93\ \text{W/m}^2$$

$$q_{eps,20\ (40°)} = 4.262 \times (40 - 20) = 85.24\ \text{W/m}^2$$

$$q_{eps,20\ (45°)} = 4.262 \times (45 - 20) = 106.55\ \text{W/m}^2$$

The calculation example for one variant of the third heated floor layout, with a vinyl panel floor without a dry screed, reinforced with an adhesive-mesh carrier layer, and without aluminium lamellas, provides the basis for calculating the remaining variants of the third multi-variant layer layout. The set of mathematical formulas adopted in the example variant was applied to calculations for other heated floor variants, each with an identical carrier layer composed of deformable cement adhesive reinforced with fibreglass mesh, for various cover floorings. The following seven floorings were assumed for underfloor heating:

- PVC flooring with integrated XPO underlay
- three-layer wooden board + polyurethane adhesive
- laminate panels + smart underlay
- two-component microcement Microflex
- PVC flooring + Eurocol 640 Eurostar adhesive
- carpet flooring + Eurocol 540 Eurosafe Special adhesive
- flexible resin flooring

The results of calculations of the thermal power of the third-layer layout of a heated floor without screeds and without aluminium heat-dissipating lamellas are summarised in Table 5.5. To compare the thermal power values of these eight floor variants, Table 5.5 includes the results calculated for a floor made of ceramic tiles from Chapter 4, with coil spacings of 12.5 cm only for a 3-cm-thick EPS insulation board and 15 and 20 cm for a 4-cm-thick XPS insulation board. Slight differences in thermal power and layer thicknesses when using EPS and XPS boards will not be significant in their comparison.

TABLE 5.5
Design Values of Thermal Power for Eight Floors of the Third Heated Floor Layout without Screeds and without Aluminium Lamellas [2]

		Design Thermal Efficiency, q_{obw} [W/m^2]			
		Coil Supply Temperature [°C]			
Coil Spacing [cm]		28	35	40	45
	PVC Floor Panels – Vinyl + Polyurethane Adhesive PU 560				
XPS	10	63.54	119.15	158.86	198.58
	12.5	53.75	100.78	134.38	167.97
	15	45.69	85.67	114.23	142.78
	20	34.22	64.17	85.55	106.94
EPS	10	63.41	118.89	158.52	198.15
	12.5	53.60	100.50	134.00	167.50
	15	45.54	85.39	113.86	142.32
	20	34.10	63.93	85.24	106.55
	Ceramic Tiles + Deformable Cement Adhesive C2S1 from Chapter 4				
EPS	12.5	65.5	122.9	163.8	204.8
XPS	15	59.3	111.2	148.2	185.3
	20	46.9	88.0	117.3	146.6
	PC Floor Panels – Vinyl with Integrated XPO Underlay				
XPS	10	64.10	120.18	160.25	200.31
	12.5	54.37	101.94	135.92	169.89
	15	46.30	86.81	115.74	144.68
	20	34.74	65.14	86.85	108.56
EPS	10	63.96	119.92	159.89	199.87
	12.5	54.21	101.64	135.52	169.40
	15	46.14	86.52	115.36	144.19
	20	34.61	64.89	86.52	108.14
	Three-Layer Wooden Board + Polyurethane Adhesive PU 560				
XPS	10	66.59	124.85	166.47	208.09
	12.5	57.19	107.24	142.98	178.73
	15	49.14	92.14	122.85	153.56
	20	37.22	69.78	93.04	116.30
EPS	10	66.43	124.56	166.08	207.59
	12.5	57.01	106.90	142.53	178.16
	15	48.95	91.79	122.38	152.98
	20	37.05	69.47	92.63	115.79
	Arendal Laminate Flooring + Smart Underlay				
XPS	10	69.03	129.44	172.58	215.73
	12.5	60.08	112.66	150.21	187.76
	15	52.14	97.76	130.35	162.94
	20	39.94	74.89	99.85	124.82
EPS	10	68.86	129.11	172.14	215.18
	12.5	59.87	112.26	149.68	187.10

(*Continued*)

TABLE 5.5 (*Continued*)
Design Values of Thermal Power for Eight Floors of the Third Heated Floor Layout without Screeds and without Aluminium Lamellas [2]

		Design Thermal Efficiency, q_{obw} [W/m^2]			
		Coil Supply Temperature [°C]			
Coil Spacing [cm]		28	35	40	45
	15	51.92	97.34	129.79	162.24
	20	39.73	74.50	99.33	124.17
Microflex Two-Component Microcement					
XPS	10	62.56	117.30	156.40	195.50
	12.5	52.67	98.76	131.69	164.61
	15	44.64	83.69	111.59	139.49
	20	33.33	62.50	83.33	104.16
EPS	10	62.43	117.06	156.07	195.09
	12.5	52.53	98.50	131.33	164.16
	15	44.50	83.43	111.25	139.06
	20	33.22	62.28	83.04	103.80
Eternal PVC Flooring + Eurocol 640 Eurostar Special Adhesive					
XPS	10	63.35	118.79	158.39	197.98
	12.5	53.54	100.39	133.86	167.32
	15	45.49	85.29	113.71	142.14
	20	34.05	63.84	85.12	106.40
EPS	10	63.22	118.53	158.05	197.56
	12.5	53.39	100.11	133.48	166.85
	15	45.34	85.01	113.35	141.68
	20	33.92	63.61	84.81	106.01
Flotex Colour Carpet + Eurocol 640 Eurosafe Special Adhesive					
XPS	10	66.60	124.87	166.50	208.12
	12.5	57.21	107.26	143.01	178.77
	15	49.15	92.16	122.88	153.60
	20	37.23	69.80	93.07	116.34
EPS	10	66.44	124.58	166.10	207.63
	12.5	57.02	106.92	142.56	178.20
	15	48.97	91.81	122.41	153.02
	20	37.06	69.49	92.66	115.82
Sikafloor Elastic Resin Floor – 300 N + Sikafloor 156 Primer					
XPS	10	63.30	118.69	158.25	197.82
	12.5	53.48	100.28	133.71	167.14
	15	45.43	85.18	113.57	141.96
	20	34.00	63.75	85.00	106.25
EPS	10	63.17	118.44	157.91	197.39
	12.5	53.34	100.00	133.34	166.67
	15	45.28	84.90	113.21	141.51
	20	33.88	63.52	84.69	105.86

5.2 RESULTS ANALYSIS

Three configurations of the lightweight underfloor heating layouts with thin screeds (layouts one and two) and without screeds (layout three) were subjected to theoretical calculations to determine the heat flux density (thermal power). Additionally, the maximum medium temperature in the coil was determined for the first-layer configuration with lamellas.

These multi-variant layouts included the use of eight floorings, three thin dry screeds or a screed-free supporting layer, four different coil spacings from 10 to 20 cm and coil supply temperatures from 28°C to 45°C, a layout with and without aluminium lamellas, and the use of two types of thermal insulation most commonly used in underfloor heating. This multi-variant layout allows for both global and detailed analysis and assessment of the calculated thermal output across the three-floor-layer configurations. The overall assessment indicates that the use of heat-dissipating metal lamellas clearly increases thermal output. Similarly, glueing the carrier layer to the thermal insulation eliminates air voids, reduces heat transfer resistance, and consequently increases thermal output. Furthermore, Table 5.5 shows that the ceramic tile floor calculated in Chapter 4 has a higher thermal output than the flooring used in Chapter 5. A detailed evaluation of the comparison results was based on an analysis of the effect of individual heated floor layers on thermal output. The detailed evaluation criteria were based on the coil temperature most commonly encountered in LRHS, namely, not exceeding 35°C, when XPS insulation boards are used, and coil spacings of 10 and 12.5 cm when metal lamellas are absent (these are proven spacings for comfortable floor use) and 12.5 cm when metal lamellas are installed, to compare the thermal outputs of layouts with and without lamellas.

5.2.1 THE INFLUENCE OF COVER FLOORING ON THE FLOOR'S THERMAL POWER

The flooring in a heated floor's layer system, specifically its technical specifications (thickness and thermal conductivity coefficient), has a certain, albeit non-deterministic, influence on the thermal output. This can best be assessed by comparing one of the three flooring layouts and a coil temperature of up to 35°C without the use of heat-dissipating lamellas, which, according to the overall assessment, have a significant impact on thermal efficiency. Comparing the thermal output data from Table 5.5 (layout without screed and without lamellas), the extreme (lowest and highest) values are highlighted in bold or white numbers on a black background, depending on the flooring type, in the table below.

To determine the percentage difference, i.e., by what percentage does the maximum thermal output for a given floor exceed the minimum for a given coil spacing, the following formula (5.1) is used:

$$\text{Increase } \% = \frac{x_{max} - x_{min}}{x_{min}} \cdot 100\% \tag{5.1}$$

where

x_{max} – maximum thermal power

x_{min} – minimum thermal power

The percentage differences relative to the maximum values (excluding ceramic flooring, which was analysed in Chapter 4) are given in the table below. The maximum values have a 0.00% increase.

		Design Thermal Efficiency, q_{obw} [W/m²]			
		Coil Supply Temperature [°C]			
Coil Spacing [cm]		28	Difference % to Max. Value	35	Difference % to Max. Value
PVC Floor Panels + Polyurethane Adhesive					
XPS	10	63.54	8.64%	119.15	8.64%
	12.5	53.75	11.78%	100.78	11.79%
Ceramic Tiles + Deformable Cement Adhesive C2S1 from Chapter 4					
EPS	12.5	65.5	0.00%	122.9	0.00%
PVC Floor Panels with Integrated XPO Underlay					
XPS	10	64.10	7.69%	120.18	7.71%
	12.5	54.37	10.50%	101.94	10.52%
Three-Layer Wooden Board + Polyurethane Adhesive					
XPS	10	66.59	3.66%	124.85	3.68%
	12.5	57.19	5.05%	107.24	5.05%
Laminated Panels + Smart Underlay					
XPS	10	**69.03**	**0.00%**	**129.44**	**0.00%**
	12.5	60.08	0.00%	112.66	0.00%
Two-Component Microcement Microflex					
XPS	10	**62.56**	**10.34%**	**117.30**	**10.35%**
	12.5	52.67	14.07%	98.76	14.07%
PVC Flooring + Eurocol 640 Eurostar Adhesive					
XPS	10	63.35	8.97%	118.79	8.97%
	12.5	53.54	12.22%	100.39	12.22%
Cover Carpet + Eurocol 540 Eurosafe Special Adhesive					
XPS	10	66.60	3.65%	124.87	3.66%
	12.5	57.21	5.02%	107.26	5.03%
Flexible Resin Flooring					
XPS	10	63.30	9.05%	118.69	9.06%
	12.5	53.48	12.34%	100.28	12.35%

The assessment of the cover flooring's impact on the floor thermal output was based on the own criteria and engineering experience. Thermal output values above 15% were considered significant, while values below this value were considered insignificant. Based on these criteria, it was concluded that

- Ceramic tile floors have the highest thermal output.
- Ignoring ceramic tile floors, laminate flooring with an underlay has the highest thermal output.
- Microcement floors have the lowest thermal output.

- At supply temperatures of both 28°C and 35°C, the largest thermal output differences occur between stoneware and microcement floors, which are approximately 24% for both temperatures, which is a significant difference.
- Ignoring stoneware floors, the largest thermal output differences occur between laminate flooring and microcement floors, which are approximately 10%–14% for both temperatures, depending on the coil spacing, which is not a significant difference.
- The differences in thermal power between the remaining floors are lower than 13%. Therefore, it can be assumed that the influence of different types of other floors on the thermal efficiency of the lightweight heated floor is not significant.

Interestingly, relatively high calculated thermal efficiency values were obtained for laminate panels and three-layer floorboards glued to the subfloor. These are characterised by lower thermal conductivity coefficients among the surveyed cover floorings (not including carpets) and the greatest thickness among all floors, excluding ceramic flooring. This confirms the complexity of calculating thermal output in lightweight floors without screeds, and the significant impact of the three equivalent thermal resistances, R_{HC}, R_i, and R_e, on the thermal exchange process. An important conclusion drawn from the calculations is that selecting the appropriate flooring layer should be considered an integral part of the heating system design process, not merely an aesthetic or functional decision. The results presented in Tables 5.3–5.5 facilitate this choice and can be used by designers and installers of lightweight underfloor heating systems for various flooring types. Furthermore, researchers can conduct experimental studies on models of lightweight heated floors, comparing the calculated thermal output values with those obtained during testing.

5.2.2 The Influence of Dry Screeds on the Floor Thermal Power

The use of different screeds in a lightweight heated floor system, particularly their technical specifications (thickness and thermal conductivity coefficient), significantly impacts the thermal output. This is best assessed by comparing one of the three floor layouts and a coil temperature of 35°C without the use of heat-dissipating lamellas, which, according to the overall assessment, have a significant impact on thermal efficiency. For screeds, the thermal output data from Table 5.4 (a layout with three screeds, without lamellas) were compared, where the extreme (lowest and highest) power values, in the table below, are highlighted in bold or white numbers on a black background, depending on the type of screed with XPS insulation for a given floor type. For comparison, a coil spacing of up to 12.5 cm was assumed, as the maximum that meets the user comfort requirements without the use of metal lamellas, as confirmed in Chapter 4.

To determine the percentage difference, i.e., how much the maximum thermal output for a screed exceeds the minimum for a given coil spacing, formula (5.1) was used:

The percentage differences in each screed, compared to the maximum value, are given in the table below. The maximum values have a 0.00% increase.

		Design Thermal Efficiency, q_{ob} [W/m²]					
		KNAUF BRIO		BAUNORM FIBERCEMENT		KNAUF DUAL FLOOR	
		Coil Supply Temperature [°C]					
Coil Spacing [cm]		35	Difference % to Max. Value	35	Difference % to Max. Value	35	Difference % to Max. Value
		PVC Floor Panels + Polyurethane Adhesive					
XPS	10	111.79	8.02%	89.13	10.28%	121.62	6.73%
	12.5	92.91	10.40%	71.17	11.77%	103.4	9.23%
		PVC Floor Panels with Integrated XPO Underlay					
XPS	10	112.64	7.20%	89.96	9.26%	122.41	6.05%
	12.5	93.79	9.36%	71.91	10.62%	104.44	8.29%
		Three-Layer Wooden Board + Polyurethane Adhesive					
XPS	10	116.61	3.55%	93.94	4.63%	126.06	2.97%
	12.5	98.01	4.65%	75.51	5.35%	108.64	4.11%
		Laminated Panels + Smart Underlay					
XPS	10	**120.75**	**0.00%**	**98.29**	**0.00%**	**129.81**	**0.00%**
	12.5	102.57	0.00%	79.55	0.00%	113.10	0.00%
		Two-Component Microcement Microflex					
XPS	10	**110.31**	**9.46%**	**87.70**	**12.08%**	**120.23**	**7.97%**
	12.5	91.38	12.25%	69.91	13.79%	101.99	10.89%
		PVC Flooring + Eurocol 640 Eurostar Adhesive					
XPS	10	111.51	8.29%	88.85	10.62%	121.35	6.97%
	12.5	92.61	10.75%	70.92	12.17%	103.24	9.55%
		Cover Carpet + Eurocol 540 Eurosafe Special Adhesive					
XPS	10	116.62	3.54%	93.95	4.62%	126.08	2.96%
	12.5	98.03	4.63%	75.52	5.34%	108.66	4.09%
		Flexible Resin Flooring					
XPS	10	111.43	8.36%	88.77	10.72%	121.27	7.04%
	12.5	92.52	10.86%	70.85	12.28%	103.15	9.65%

The assessment of the effect of dry screeds on floor thermal output was based on the own criteria and engineering experience. It was assumed that thermal output values, similar to those for cover flooring, are significant above 15% and insignificant below this value. Based on these criteria, it was concluded that:

- KNAUF DUAL FLOOR screeds with laminate flooring have the highest thermal output at a 10-cm coil spacing – 129.81 W/m².
- BAUNORM FIBERCEMENT screeds with microcement flooring have the lowest thermal output at a 10-cm spacing – 87.70 W/m².
- Therefore, the maximum percentage difference in thermal output at a 10-cm spacing and using different flooring and screeds was 48%, according to formula (5.1), which is a significant difference.

- At a supply temperature of 35°C, the largest differences in thermal output for the selected screed and different flooring types occurred for the BAUNORM FIBERCEMENT screed, amounting to 13.79%, which is not a significant difference.
- The differences in thermal output between other independently analysed screeds on different floorings are less than 13%, which means that the impact of different types of floorings on the thermal efficiency of the lightweight heated floor is insignificant.
- The difference in thermal output between different screeds using the same flooring is significant. For laminated panels, the difference between the KNAUF DUAL FLOOR screed and BAUNORM FIBERCEMENT is 32%–42%, depending on the coil spacing, or 37%–46% for microcement flooring.

An important conclusion from the calculations is that selecting the appropriate screed layer should be an integral part of the heating system design process, especially since different screed types significantly affect the thermal efficiency of the heated floor. Designers and installers of lightweight underfloor heating systems with different screed types can use the results. Similarly, as with cover flooring, it was noted that a relatively high calculated thermal efficiency was achieved for the KNAUF DUAL FLOOR screed, despite its similar thermal conductivity coefficient to other screeds, but its greatest thickness compared to other screeds. This fact also confirms the complexity of calculating thermal output in lightweight floors without screeds. Furthermore, an additional advantage of the KNAUF DUAL FLOOR screed is its lowest purchase cost of the screeds analysed. However, its installation is more complex, requiring the joining of two independent 12.5-mm-thick boards.

5.2.3 The Influence of Aluminium Lamellas on the Floor Thermal Power

In one of the heated floor layers with screeds, aluminium lamellas were used to increase thermal output and evenly distribute heat. The thickness of the aluminium lamellas is the smallest among heated floor layers. At the same time, their thermal conductivity, with a coefficient of $\lambda = 200$ W/m×K, is approximately 800 times higher than that of other materials. This significant difference in thermal conductivity significantly impacts the thermal output of the heated floor. This can best be appreciated by comparing three-floor-layer layouts with a coil supply temperature of 35°C and a distance between heating pipes of 12.5 cm – the first with different screeds using aluminium lamellas, the second with screeds but no lamellas, and the third without screeds or lamellas. When comparing the thermal output data from Tables 5.3–5.5 (for the layout with screeds and lamellas, the layout with screeds and no lamellas, and the layout without screeds, respectively), the extreme values are highlighted in the newly developed table below, using bold or white numbers on a black background depending on the flooring type. A 12.5-cm distance between heating pipes was assumed for the comparison of these three flooring layouts, as this is a comfortable distance for the layout without metal lamellas in both the central and edge zones of the underfloor heating system, in accordance with the guidelines in paragraph 4.

To determine the percentage difference, i.e., how much the maximum thermal output using lamellas exceeds the minimum for a given coil spacing, formula (5.1) was used.

These percentage differences compared to the maximum values are given in the table below, where 12.5no denotes the coil spacing on a floor with screed without lamellas, 12.5la with lamellas, and 12.5wz without screed and without lamellas. The maximum values have a 0.00% increase.

		Design Thermal Efficiency, q_{ob} [W/m²]					
		KNAUF BRIO		BAUNORM FIBERCEMENT		KNAUF DUAL FLOOR	
		Coil Supply Temperature [°C]					
Coil Spacing [cm]		35	Difference % to Max. Value	35	Difference % to Max. Value	35	Difference % to Max. Value
		PVC Floor Panels + Polyurethane Adhesive					
XPS	12.5 no	92.91	54.41%	71.17	100.52%	103.54	39.26%
	12.5 la	138.75	3.39%	137.42	3.85%	140.01	2.99%
	12.5 wz	100.78	42.35%	100.78	41.72%	100.78	43.07%
	wzm						
		PVC Floor Panels with Integrated XPO Underlay					
XPS	12.5 no	93.79	52.96%	71.91	98.46%	104.44	38.06%
	12.5 la	139.20	3.06%	137.93	3.47%	140.40	2.70%
	12.5 wz	101.94	40.73%	101.94	39.99%	101.94	41.45%
		Three-Layer Wooden Board + Polyurethane Adhesive					
XPS	12.5 no	98.01	46.37%	75.51	88.99%	108.64	32.72%
	12.5 la	141.28	1.54%	140.28	1.73%	142.24	1.37%
	12.5 wz	107.24	33.77%	107.24	33.08%	107.24	34.46%
		Laminated Panels + Smart Underlay					
XPS	12.5 no	**102.57**	**39.87%**	**79.55**	**79.40%**	**113.10**	**27.49%**
	12.5 la	143.46	0.00%	142.71	0.00%	144.19	0.00%
	12.5 wz	112.66	27.34%	112.66	26.67%	112.66	27.99%
		Two-Component Microcement Microflex					
XPS	12.5 no	**91.38**	**56.99%**	**69.91**	**104.13%**	**101.99**	**41.38%**
	12.5 la	137.97	3.98%	136.52	4.53%	139.34	3.48%
	12.5 wz	98.76	45.26%	98.76	44.50%	98.76	46.00%
		PVC Flooring + Eurocol 640 Eurostar Adhesive					
XPS	12.5 no	92.61	54.91%	70.92	101.23%	103.24	39.66%
	12.5 la	138.60	3.51%	137.24	3.99%	139.88	3.08%
	12.5 wz	100.39	42.90%	100.39	42.16%	100.39	43.63%
		Cover Carpet + Eurocol 540 Eurosafe Special Adhesive					
XPS	12.5 no	98.03	46.34%	75.52	88.97%	108.66	32.70%
	12.5 la	141.29	1.54%	140.29	1.72%	142.25	1.36%
	12.5 wz	107.26	33.75%	107.26	33.05%	107.26	34.43%
		Flexible Resin Flooring					
XPS	12.5 no	92.52	55.06%	70.85	101.43%	103.15	39.79%
	12.5 la	138.56	3.54%	137.19	4.02%	139.85	3.10%
	12.5 wz	100.28	43.06%	100.28	42.31%	100.28	43.79%

The assessment of the effect of aluminium lamellas on the floor's thermal output was based on the own criteria and engineering experience. It was assumed that thermal output values, similar to those for flooring and screed, are significant above 15% and insignificant below that value. Based on these criteria, it was concluded that

- Floors achieved the highest thermal output at a 12.5-cm coil spacing with lamellas placed under KNAUF DUAL FLOOR screed with laminate flooring – 144.19 W/m^2.
- The lowest thermal output at a 12.5-cm spacing was achieved by floors without lamellas, with BAUNORM FIBERCEMENT screed and microcement flooring – 69.91 W/m^2.
- Therefore, the maximum percentage difference in thermal output at a 12.5-cm spacing and using a floor layout with screeds, with and without lamellas, was as much as 104.13%, according to formula (5.1), which is a very significant difference.
- Comparing a floor with lamellas and screed to a floor without screed and lamellas, the thermal output differences were also significant, ranging from 26.67% to 46.00%. However, there is not as much difference here as in the case of using a floor with screeds and without lamellas.
- At a flow temperature of 35°C, the different types of screed or cover flooring do not play a significant role in a floor layout with lamellas. Then, the thermal power differences between screed types vary up to 4.53%.
- The maximum thermal power of the floor layout without screed is 112.66 W/m^2, and with KNAUF DUAL FLOOR screed without lamellas, it is 113.1 W/m^2 with laminate flooring. The difference in this case is imperceptible. However, comparing the minimum thermal power values of these floor layouts, we obtain 98.76 W/m^2 for the layout without screed and only 69.91 W/m^2 for the layout with screed without lamellas, a significant difference of 41%.

An important conclusion drawn from the calculations is that the use of metal lamellas significantly affects the thermal efficiency of the heated floor. Comparing the calculated thermal output of the floor with and without lamellas shows that structures equipped with dissipative elements transfer higher heat flux to the room. These differences are visible when comparing specific floor configurations with the same screed type, flooring, supply temperature, and coil spacing. The table above applies only to a 12.5-cm coil spacing, which ensures comfortable use regardless of whether lamellas are used. If this is not the case and the coil spacing exceeds 12.5 cm, it is worth considering that effective heat transfer may be uneven, creating the so-called cold spots that result in reduced comfort. The values and percentages provided in the tables facilitate the selection of the appropriate solution for the type of flooring and screed, which designers and installers of lightweight underfloor heating systems can use.

5.2.4 Influence of Coil Spacing and Supply Temperature on the Floor Heating Power

Each of the presented Tables 5.3–5.5, and the comparative tables, which specify the percentage differences in thermal output, clearly demonstrates the significant impact that the coil spacing has on the thermal efficiency of the heated floor in each system. Maximum thermal outputs were achieved at a distance of 10 cm. Increasing the distance between the heating pipes results in a local decrease in floor surface temperature, depending on the assumed distance, thereby reducing the heat flux density transferred to the room. Similarly, the heating medium supply temperature significantly affects thermal efficiency. Figures 5.2–5.4 show examples of linear

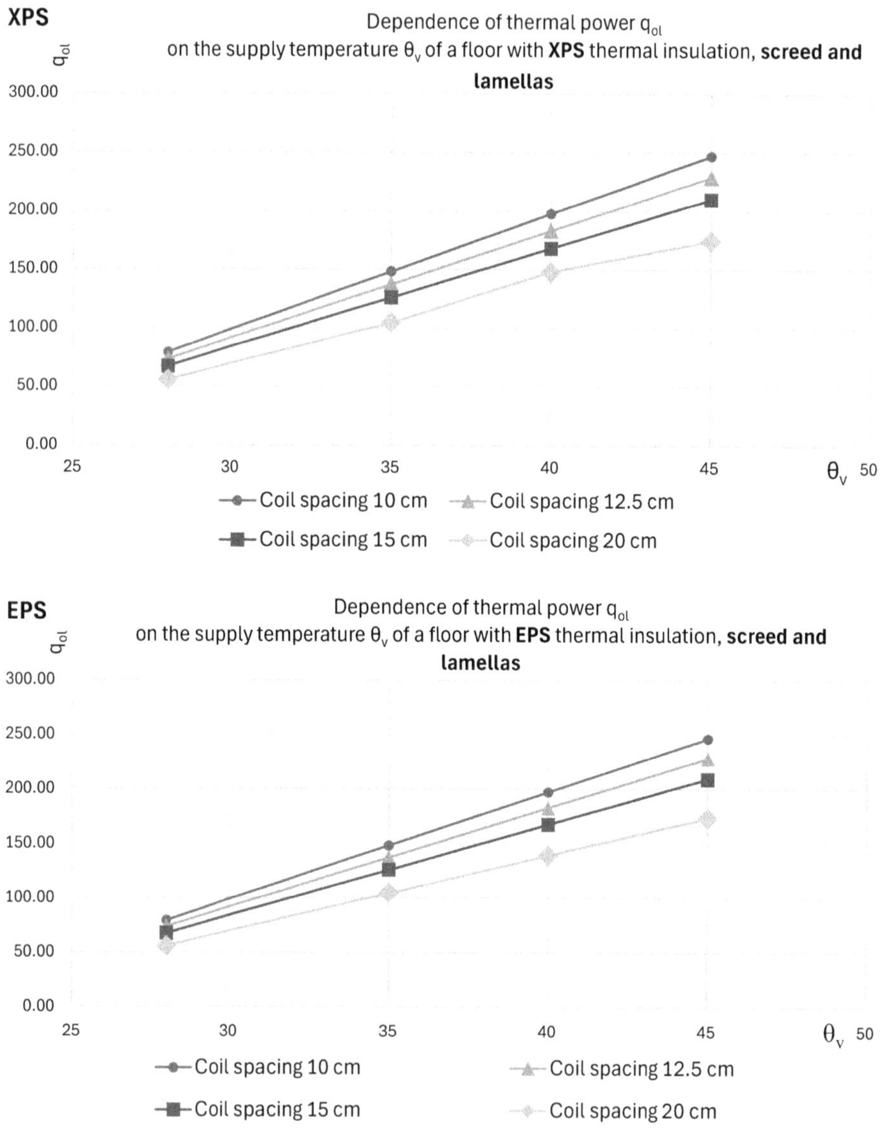

FIGURE 5.2 Example graph of the dependence of thermal power on the supply temperature of a floor with EPS and XPS thermal insulation, **screed and lamellas**, with different coil spacings.

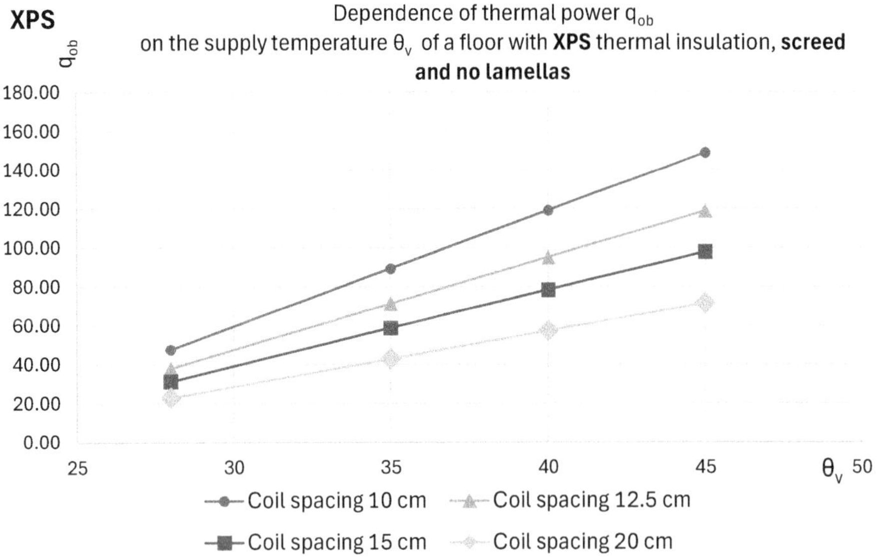

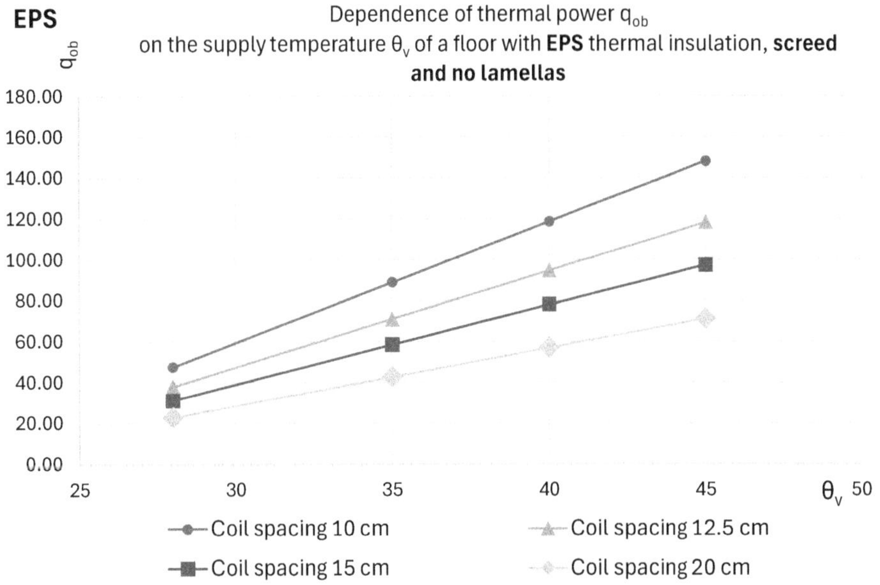

FIGURE 5.3 Example graph of the dependence of thermal power on the supply temperature of a floor with EPS and XPS thermal insulation, **screed and no lamellas**, with different coil spacings.

dependencies of thermal output, based on Tables 5.3–5.5, for floor layouts with and without lamellas, and with and without screeds, as a function of supply temperature.

The calculated thermal power of the floor, shown in the three figures above, is linear regardless of the floor layout. Therefore, it can be assumed that the assumed

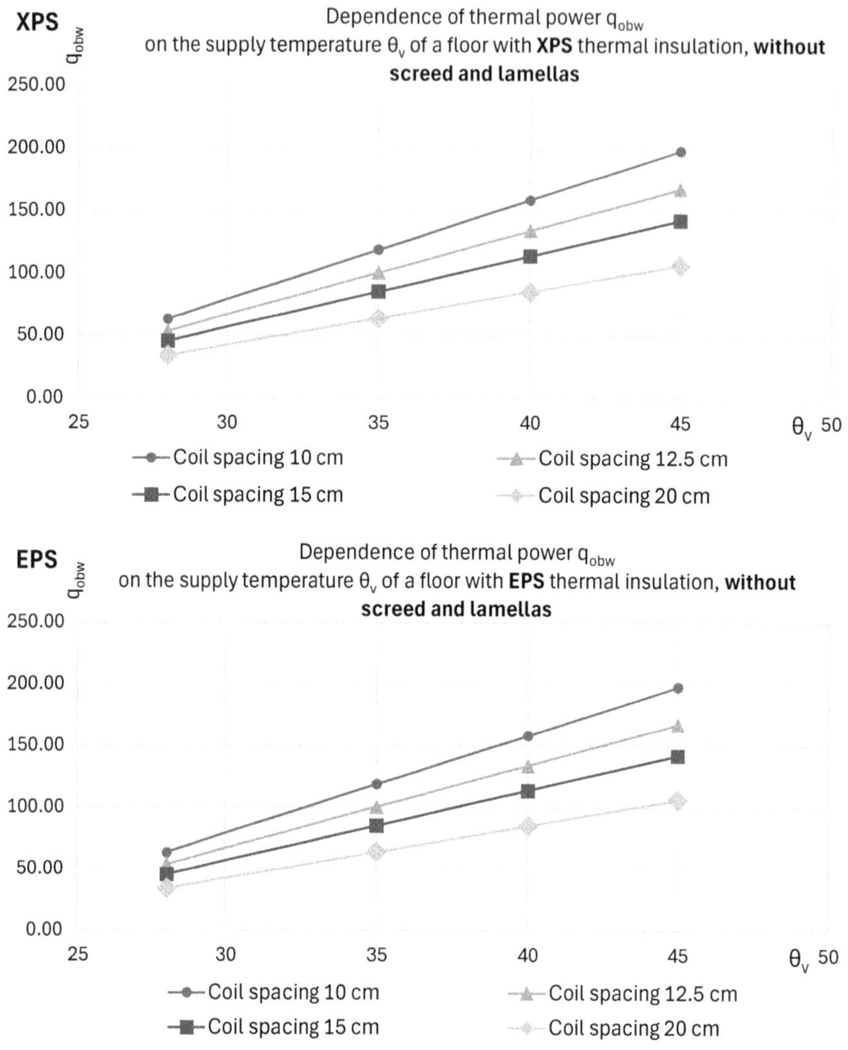

FIGURE 5.4 Example graph of the dependence of thermal power on the supply temperature of a floor with EPS and XPS thermal insulation, **without screed and lamellas**, with different coil spacings.

multivariate floor layouts with different flooring, screeds, lamellas, or without them, and with different spacings and supply temperatures, have the same linear thermal power. This is important when interpolating the available data with the required supply temperature for heating pipes. This aspect is worth considering when selecting low-temperature power sources for lightweight underfloor heating systems, such as heat pumps, where maintaining the lowest possible heating medium temperature is recommended. Lower supply temperatures, which exhibit a linear relationship, can be achieved by minimising the distance between the heating pipes and vice versa. Smaller coil spacing and a lower heating medium temperature result in lower operating costs. The higher cost of heating pipes, due to their density, is not a significant investment cost.

5.2.5 The Influence of Thermal Insulation on the Floor Thermal Power

Tables 5.3–5.5, based on which the percentage differences in thermal output are determined, show that the thermal insulation used does not significantly affect thermal output. This is because the technical parameters of these insulations are similar, and their thicknesses were not varied. To compare the impact of thermal insulation on thermal efficiency, it would be necessary to use insulation with significantly different technical parameters and thicknesses and assume heat loss from the bottom of the room due to temperature differences. The principles of building physics clearly indicate that thicker thermal insulation, smaller temperature differences between rooms, and a lower thermal conductivity coefficient reduce heat loss. The slightly better thermal properties of XPS boards compared to EPS slightly increase the thermal output of the heated floor. However, this is not a significant value. However, insulation boards should be appreciated from another perspective. The calculations assumed XPS boards with a compressive strength of 300 kPa, while EPS boards had a strength of 200 kPa. This means that XPS boards have an additional advantage: they can withstand higher design loads on a heated floor. This is particularly important when we do not use a load-bearing screed layer, but an adhesive-mesh reinforcement layer. This type of reinforcement was analysed, and the stability of the floor system without screeds was confirmed, provided the insulation slabs had a compressive strength of not less than 200 kPa. Therefore, caution should be exercised when using lower-strength EPS boards in an LRHS without a screed, bearing in mind that thermal stresses in screedless floors directly affect the adhesive-mesh reinforcement layer, which must be carefully designed and installed in accordance with the LRHS manufacturer's guidelines.

5.3 SUMMARY

Calculations and multi-criteria analysis of the thermal output of the heated floor were conducted in three layouts. Each floor had eight different floor coverings, four coil spacings – 10, 12.5, 15, and 20 cm – supplied with water at four temperatures: 28°C, 35°C, 30°C, and 45°C, and two types of thermal insulation: XPS and EPS. The three layouts differed in that the first layout consisted of three different types of dry screeds: BAUNORM FIBERCEMENT, KNAUF BRIO, and KNAUF DUAL FLOOR, with thicknesses of 10, 18, and 25 mm, respectively, resting on heat-dissipating aluminium lamellas. The second layout was equipped with the same dry screeds but without the aluminium lamellas layer. The third layout had no screeds or lamellas. Instead of screeds, the load-bearing element of the heated floor was an adhesive-mesh layer, 3 mm thick, composed of a deformable cement adhesive mortar type C2S1 reinforced with glass fibre mesh, covered with a 5-mm-thick layer of

SCHOMBURG SOLOPLAN-30 self-levelling screed. For these three floor layouts, with variations in cover flooring, coil spacing and temperature, and thermal insulation type, theoretical heat flux density calculations were performed and presented in tables. Based on these tables, a percentage comparison of differences in thermal output and the impact of each layer and floor equipment on the performance of the heated floor, with and without screeds, was made. Based on this, and assuming the established evaluation criteria from the analysis of the results, the most important conclusions can be drawn:

- The influence of cover flooring on the thermal efficiency of a heated floor can be considered less significant, as the differences in thermal power between cover floorings (excluding the ceramic tile floor analysed in Chapter 4) are lower than 15%.
- The thermal output of a floor depends not only on the thermal conductivity coefficient of the material but also on its thickness, in which heat is stored. This can be confirmed by comparing the higher thermal output of a floor covered with thicker laminate panels with a lower thermal conductivity coefficient to that of thinner floors with higher thermal conductivity. Furthermore, laminate panels have another important advantage: they are one of the cheapest materials used in calculations and are easy to install.
- Using different types of screeds significantly affects the thermal efficiency of a heated floor, even with the same cover flooring. This is confirmed by the use of laminated panels, where the thermal output differs between KNAUF DUAL FLOOR and BAUNORM FIBERCEMENT screeds by 32%–42%, depending on the coil spacing. With microcement flooring, the difference is even greater, ranging from 37% to 46%.
- Using aluminium lamellas significantly affects the thermal efficiency of a heated floor. Comparing the calculated thermal output of a floor with and without lamellas shows that structures equipped with dissipative elements achieve higher heat flux values (up to twice) transferred to the room.
- For various types of heating floors, and assuming a floor without metal lamellas, a better solution in terms of thermal efficiency will be a system that eliminates the screed and replaces it with an adhesive-mesh layer.
- The distance and temperature of the coil supply are very important factors in designing the ability to regulate the thermal efficiency of the heated floor in any layout.
- The calculated thermal output of the floor relative to the supply temperature and the distance between the heating pipes is linear, regardless of the floor layout, which is helpful when designing the floor using data interpolation.
- The thermal insulation materials used, such as XPS and EPS boards, do not significantly impact the thermal output results, regardless of the floor layout. To conduct a more in-depth analysis of the impact of thermal insulation on the thermal efficiency of a heated floor, materials with significantly greater thermal conductivity differences, different thicknesses, and different temperatures between adjacent rooms would need to be used.

- Based on the calculation of the maximum permissible medium temperature in the heating pipe of the first example layout of a heated floor covered with vinyl panels with a dry screed and metal lamellas θ_H^{max}, it should be assumed that its safe operating value should not exceed 51°C. This value is 9°C higher than the maximum temperature calculated in Chapter 4 for a ceramic tile floor without screed. A higher permissible supply temperature in a heated floor system with screeds may be characterised by lower thermal efficiency at the same supply parameters and coil distance relative to the floor system without screeds.

Based on the obtained thermal output results for the three floor layouts, it was demonstrated that the highest thermal efficiency was achieved with lightweight underfloor heating systems equipped with heat-dissipating lamellas, placed on an XPS insulating underlayment, with coil spacing no greater than 10–12.5 cm. The best thermal output was achieved with Knauf Dual Floor dry screed with aluminium lamellas, and a floor finished with laminated panels installed on a system underlayment.

I would recommend that experimental studies, in the laboratory or "in situ", on the lightweight underfloor heating systems selected in this chapter confirm the reliability of theoretical thermal power calculations. Experimental studies would also determine the temperature distribution across the various floor surfaces, thereby contributing to the definition of comfortable operating conditions. Future studies of the LRHS without metal lamellas selected in this monograph will enable verification of the theoretical thermal power calculations used by Karpiesiuk [34]. It's also worthwhile to expand experimental research to examine cooling power, particularly in floor systems without screeds, which can block the free flow of cooling. These research directions could contribute to the universal application of lightweight floor systems for both heating and cooling. The result would be a versatile surface heating and cooling system that would reduce investment and operating costs. Furthermore, it's a low-temperature system that perfectly aligns with environmental concerns.

ABBREVIATIONS

C2S1	deformable cement adhesive mortar
EPS	expanded polystyrene
LRHS	lightweight radiant heating system
PC	polymer composite
PVC	poly vinyl chloride
XPS	extruded polystyrene

REFERENCES

1. Grand View Research. Radiant Heating System Market Size, Share & Trends Analysis Report By Product (Electric Systems, Hydronic Systems, Gas Systems) By Application, By Installation Surface, By Region, And Segment Forecasts, 2025–2033, 100 pages, October 8, 2025, ID: GVR-4-68040-767-2.

2. Zientara D.K., *Thermal efficiency of various flooring types in a lightweight floor system with thermal action.* Engineering Diploma Thesis, Faculty of Construction and Environmental Engineering, Warsaw University of Life Sciences, 2026. (in Polish).
3. Zhou W., Zhu W., Zhu Y., Li T., Comparative study on the thermal performance of dry and wet radiant floor heating under the characteristics of intermittent energy use. *Journal of Xi'an Polytechnic University*, 37 (6) (2023), pp. 26–33. https://doi.org/10.13338/j.issn.1674-649x.2023.06.004
4. KAN. Surface heating and cooling. COMFORT AND EFFICIENCY. Install the future. Access: https://en.kan-therm.com/resfile/open/2284/kan-therm-catalogue-surface-heating&cooling-en-25-12-02.pdf
5. NORDTEST Method NT VVS127: UDC 697.1: Approved 2001-11. Floor heating systems. Design and type testing of waterborne heat systems for lightweight structures. Finland.
6. Karpiesiuk J., Correction of the thermal power calculation in the lightweight floor heating based on NT VVS127:2001 standard. *Journal of Civil and Environmental Engineering*, 10 (1) (2024), pp. 001–004. https://doi.org/10.17352/2455-488X.000076
7. Werner-Juszczuk A.J., The calculations of the efficiency of the lightweight floor heating according to Nordtest method NT VVS 127. *Innovations – sustainability – modernity – openness modernity in engineering.* Edited by Dorota Anna Krawczyk, Iwona Skoczko, Ewa Szatyłowicz, Series of monographs, Volume 43. Printing House of Bialystok University of Technology, Bialystok, 2021, pp. 159–172. ISBN 978-83-66391-90-1 (eBook). https://doi.org/10.24427/978-83-66391-90-1
8. Weitzmann P., Svendsen S., Method for calculating thermal properties of lightweight floor heating panels based on an experimental setup. *International Journal of Low Energy and Sustainable Buildings*, 3 (2005), pp. 1–15.
9. Werner-Juszczuk A.J., The influence of the thickness of an aluminum radiant sheet on the performance of the lightweight floor heating. *Journal of Building Engineering*, 44 (2021), p. 102896.
10. Żukowski M., Karpiesiuk P., Test results of a radiant radiator with a very low height. *Instal*, 10 (2015), pp. 38–41. (in Polish).
11. Dongliang Z., Ning C., Zijie W., Experimental and numerical analysis of lightweight radiant floor heating system. *Energy and Buildings*, 61 (2013), pp. 260–266.
12. Kim K.W., Olesen B., Radiant heating and cooling systems: part 1. *ASHRAE Journal*, 57 (2015), pp. 28–37.
13. Barlinek, Declaration of Performance. SPC floor panels without integrated underlayment, 2024. Access: https://www.barlinek.com.pl/wp-content/uploads/2024/09/deklaracja-wlasciwosci-uzytkowych-spc-bez-podkladu-zintegrowanego.pdf
14. Murexin, Manual: PU 560 Polyurethane Parquet Adhesive, 2023. Access: https://www.murexin.pl/app/media/PL/technicaldatasheets/KT%20PU%20560%20[66024_PARKETTKLEBSTOFF%20PU%20560_20230706_085906].pdf
15. Barlinek, Declaration of Performance. SPC floor panels with integrated underlayment, 2024. Access: https://www.barlinek.com.pl/wp-content/uploads/2024/09/deklaracja-wlasciwosci-uzytkowych-spc-z-podkladem-zintegrowanym.pdf
16. Barlinek, Declaration of Performance. Three-layer floorboard, 2018. Access: https://www.barlinek.com.pl/wp-content/uploads/2018/03/deklaracja-wlasciwosci-uzytkowych-deska-trojwarstwowa.pdf
17. Pergo, Technical Data Sheet. Original Laminate Arendal, 2025. Access: https://int.pergo.com/-/media/imported%20assets/flooring/8/7/0/pgtdslmprowarendalen-pdf224709.ashx?rev=223645723ac74dd58bf797191a0f0005&type=original&filename=Technica l+Data+Sheet.pdf
18. Pergo, Technical data sheet. SmartUnderlay, 2016. Access: https://share.google/Xi68GxJYRk1SkNclg

19. Cemher, MicroFlex. Technical datasheet, 2023. Access: https://shop.medusa-design.com/wp-content/uploads/2024/03/MicroFlex_TehnicniList_Cemher_MedusaDesign.pdf
20. Forbo, Technical specifications. Forbo Eternal, 2025. Access: https://forbo.blob.core.windows.net/forbodocuments/5eb92f81-1e10-4df8-9512-7ddfe323d946/Forbo%20Eternal%20specyfikacja%20techniczna.pdf
21. Forbo, Technical specifications. Forbo Eurocol 640 Eurosafe Special, 2021. Access: https://www.forbo.com/eurocol/pl/640-eurostar-special/ekxm16
22. Forbo, Technical specifications. Forbo Flotex Colour, 2024. Access: https://forbo.blob.core.windows.net/forbodocuments/557c8cc1-2e33-4cb7-a258-bf7fc58fcbfd/Forbo%20Flotex%20Colour%20specyfikacja%20techniczna.pdf
23. Sika, Product data sheet. Sikafloor – 300 N, 2016. Access: https://www.promain.co.uk/specsheets/Sikafloor-330-N-Product-Data-Sheet.pdf?srsltid=AfmBOorSyIwk6ff3pAztGMK-d4g8mGmx1R62mEU9KR- 9bgYHEHM-WsGd
24. Sika, Declaration of Performance. Sikafloor – 156, 2018. Access: https://pol.sika.com/dms/getdocument.get/a91c6a50-9e84-34d8-bb35-e437807acce2/DoP_Sikafloor_156_60492300-DE_EN1504-2_pol.pdf
25. EN ISO 10456:2009. Building materials and products: hygrothermal properties tabulated design values and procedures for determining and declaring declared and design thermal values. European Committee for Standardization (CEN): Brussels, Belgium.
26. ISO 6946. Third edition, 2017-06. (Building components and building elements: thermal resistance and thermal transmittance: calculation methods). International Organization for Standardization, Geneva, Switzerland.
27. Muniak D., Water surface heating: wall and ceiling systems. Polski Instalator. 11-12/2022 (307), (2022), 20–24. (in Polish).
28. Suprema Eco Solutions, Manual. BAUNORM Fibercement Board, 2023. Access: https://supremaeco.com/wp-content/uploads/2025/03/KT-BAUNORM-Fibercement-Board.pdf
29. Knauf, Manual. Knauf Brio 18, 2024. Access: https://knauf.com/api/download-center/v1/assets/aa0a98ea-f6e2-4c80-b4a7-7c1fd590def2?download=true&country=pl
30. Knauf, Manual. Knauf Dual Floor, 2025. Access: https://knauf.com/api/download-center/v1/assets/aa0a98ea-f6e2-4c80-b4a7-7c1fd590def2?download=true&country=pl
31. Elektra Kardo technical catalogue, Poland, January 2023.
32. Kisan, Technical data: metal heating lamellas, 2022. Access: https://kisan.pl/wp-.content/uploads/2022/05/Kisan_Lamel_grzejny_metalowy.pdf
33. Herz, Declaration of Performance. Herz Pipefix PE-RT/AL/PE-RT 16x2 multilayer pipes, 2017. Access: https://www.katalog.herz.com.pl/img/cms/stefan/DiA/DWU-(3002b)-Rura-EN-PL.pdf
34. Karpiesiuk J., *Thermal efficiency of a lightweight surface heating system*. Białystok, GlobeEdit, 2019. International Book Market Service Ltd. Polish Edition.
35. SCHOMBURG, Safety Data Sheet. SOLOPLAN – 30. Fluidized floor mortar, 2014. Access: https://suez.b-cdn.net/media/e9/c6/fe/1677746878/SOLOPLAN-30%20kch.pdf

6 Multi-Criteria Decision Analysis (MCDA) with Cost Analysis of Selected Types of Underfloor Heating

When making a decision, taking into account various technical, economic, and other issues, it is worth ensuring it is the best possible one under the given circumstances. The choice can be made using various methods. Still, it is best if we can conduct a comparative analysis of many options with similar applications and establish firm, valid criteria for their evaluation. A comparative assessment of possible decisions in light of the criteria selected allows you to determine which of them meets these requirements to the highest degree. When deciding on a specific variant, it is worth basing the decision not only on a subjective assessment, even if supported by mathematical analysis, but also on inviting advisors from the immediate environment and experts in the given field to participate. Then, there is a good chance that our final decision will be the best possible and optimal. The need for optimisation is greater when the responsibility for the final choice of a specific variant from many options lies not only with us but also with third parties, and the consequences of the decision affect a specific group of people. The subjects of the decision-making process may be matters relating to all areas of life, including personal, social, economic, and political. The need to optimise a specific decision is related to the investment process and the principle of rational resource management. One such investment decision may be choosing the most advantageous design solution for underfloor heating in an entire newly constructed single-family building, or a part of one, when we plan renovation or modernisation to improve the comfort of use. It is then worth using multi-criteria decision analysis (MCDA) as a methodology.

The MCDA example shows the procedure for selecting the most advantageous underfloor heating system for a single-family, detached, single-story house. Additionally, the MCDA considers performing this heating only in rooms up to 20 m^2 (small areas), when we plan to improve the heating system locally and achieve the comfort of a warm floor.

The MCDA should be preceded by an analysis of the underfloor heating systems with various costs, not only of the materials but also of labour and home operation. Table 6.1 includes the average retail and wholesale costs of materials

DOI: 10.1201/9781003679004-6

TABLE 6.1
The Analysis Cost of the Underfloor Radiators with Nine Various Designs

No.	Radiator Type*	Net Retail Price (PLN/m²)**	Net Wholesale Price (PLN/m²)**
1	"Wet" screed (concrete or anhydrite screed)	210/310 (90/170 M; 120/140 L) second position – small areas	185/75 M,110 L
2	"Dry" gypsum screed (screed without lamellas)	270 (160 M, 110 L)	230 (130 M, 100 L)
3	"Dry" gypsum screed (screed with aluminium foil lamellas)	310 (200 M,110 L)	255 (155 M, 100 L)
4	"Dry" screed (cement screed with steel lamellas)	490 (370 M,120 L)	430 (320 M,110 L)
5	**Without screed** (mesh with C2S1 adhesive without lamellas)	225 (125 M, 100 L)	190 (100 M, 90 L)
6	**Without screed** (mesh with C2S2 adhesive without lamellas)	250 (150 M, 100 L)	210 (120 M, 90 L)
7	**Without screed** (polyurethane adhesive with aluminium lamellas)	290 (210 M, 80 L)	240 (165 M, 75 L)
8	"Wet" screed with electric heating cables	225 (105 M, 120 L)	190 (80 M, 110 L)
9	"Wet" screed with electric heating mats	300 (180 M, 120, L)	242 (132 M, 110 L)

M, material; **L**, labour, **where 4.0 PLN/m² ≈ 1 $/m² or 0.9 €/m²**.
*Uniform assumptions of individual systems, with their differences, include: the use of XPS boards with a thickness of 3 cm and the laying of aluminium foil, underfloor heating pipes every 10 cm. It is more expensive to lay a "wet" screed than an adhesive with mesh.
Average market prices – retail and wholesale of material producers and contractors for **2023 were assumed. In the case of polyurethane adhesive (7), the cost of material was reduced by up to 70%, and the labour cost was fixed at PLN 80/m² (for retail price) due to the faster process of laying ceramic tiles with the polyurethane adhesive. The reason was that in cement adhesives, there is a double process of installing ceramic tiles – first, a layer of cement adhesive with mesh is put in, and then adhesive with ceramic tiles is added. On the other hand, the polyurethane adhesive is laid in one step, together with the ceramic tiles.

and labour in 2023, using the same products and compatibility of assembly technologies for the territory of Poland.

The following net material prices without VAT (value-added tax) were assumed, where 4.0 PLN/m² ≈ 1 $/m² or 0.9 €/m²:

No.	Product	Net Retail Price (PLN/m²)[a]	Net Wholesale Price (PLN/m²)[a]
1	KNAUF Dual Floor 2 × 12.5 mm on WATERPANEL X with pipes	60 + 60 + 40 = 160M/110R	50 + 45 + 35 = 130M/100R
2	KNAUF Dual Floor 2 × 12.5 mm on WATERPANEL XE with pipes	60+ 100 + 40 = 200M/110R	50 + 70 + 35 = 155M/100R

(Continued)

(*Continued*)

No.	Product	Net Retail Price (PLN/m²)[a]	Net Wholesale Price (PLN/m²)[a]
3	KNAUF Aquapanel, 22 mm on WATERPANEL X with pipes and steel lamellas	220 + 60 + 40 + 50 = 370M/120R	200 + 45 + 35 + 40 = 320M/100R
4	Wet screed, XPS 3 cm Aluminium foil PE-RT/AL/PE-RT pipe 16 mm	30 + 18 + 2 + 40 = 90M/120R	25 + 14 + 1 + 35 = 75M/110R
5	WATERPANEL X with C2S1 GLUE pipes WITH MESH	60 + 40 + 15 + 10 = 125M/100R	45 + 35 + 12 + 8 = 100M/90R
6	WATERPANEL X with C2S2 GLUE pipes WITH MESH	60 + 40 + 40 + 10 = 150M/100R	45 + 35 + 32 + 8 = 120M/90R
7	WATERPANEL XE with BondT8 (70% of the cost)	100 + 40 + 70 = 210M/80R	70 + 35 + 60 = 165M/75R
8	Wet screed with 17 W/m electric heating cables on a mesh, XPS 3 cm, aluminium foil	30 + 18 + 2 + 55 = 105M/120R	25 + 14 + 1 + 40 = 80M/110R
9	Wet screed with electric heating mats in C2S1 glue, XPS 3 cm, aluminium foil	30 + 18 + 2 + 115 + 15 = 180M/120R	25 + 14 + 1 + 80 + 12 = 132M/110R

[a] Assumptions for material costs per 1 m^2 are in the table above.

Retail/wholesale prices of the material:
PLN 60/50 – KNAUF Dual Floor 2 × 12.5 mm ("dry", gypsum screed)
PLN 220/200 – KNAUF Aquapanel, 22 mm ("dry", cement screed)
PLN 60/45 – WATERPANEL X (XPS board with grooves, without lamellas) from [1]
PLN 100/70 – WATERPANEL XE (XPS board with grooves, with lamellas) from [1]
PLN 40/35 – PE-RT/AL/PE-RT heating coil 16 mm
PLN 50/40 – steel lamellas
PLN 30/25 – "wet" cement screed
PLN 18/14 – XPS thickness 3 cm
PLN 2/1 – aluminium foil
PLN 15/12 – C2S1 adhesive cement mortar
PLN 40/32 – C2S2 adhesive cement mortar
PLN 10/8 – armoured mesh, with a weight of 320 g/m^2
PLN 70/60 – polyurethane adhesive BondT8 (70% of the cost), after taking into account the ceramic tiles laying process
PLN 55/40 – electric heating cables mounted on a mesh made of [2]
PLN 115/80 – electric heating mats from [2]

When comparing item no. 1 ("heavy" underfloor heating) and no. 5 ("lightweight" underfloor heater without lamellas) from Table 6.1, it can be concluded that the prices are similar, to the detriment of the lightweight system. However, if we assume the need to create small heating surfaces, the price of traditional "wet" heating is approximately 35% higher. The reason is the higher cost of making concrete finishes on small surfaces.

6.1 IDENTIFICATION OF THE MCDA OBJECT

The MCDA concerns a house with a usable area of 100 m^2. The specific heating power demand is 47 W/m^2, with the seasonal rate of heat demand for heating E_A = 64.20 kWh/year·m^2, which, according to the Austrian guidelines described in [3], classifies the building as energy-saving, class C (energy efficiency in the range of 50–100 kWh/year·m^2). The house is intended to accommodate a family of four. The output data is based on the Elektra catalogue, edition 09/2021 and 02/2024 [2].

MCDA was adopted for eight variants of underfloor heating, including electric and water underfloor heating. Electricity was used as the power supply for electric cables and mats, or energy from natural gas (using a gas boiler) when using a water heating system with a heating coil with a diameter of 16 × 2 mm and a spacing of 10 cm. The floor is made of ceramic tiles, and its thermal resistance depends on the type of underfloor heater.

6.1.1 The Following Variants Were Taken into Account for Analysis

Variant 1 corresponds to traditional *water heating* with a heating coil placed *in a* "wet" concrete surface according to standard PN-EN 1264-2:2008 [4] or PN-EN 15377-1:2008 [5].

Variant 2 corresponds to *water heating* with a heating coil placed in thermal insulation grooves, all covered *with a "dry" gypsum screed without* heat-dissipating *lamellas*, based on the research of Karpiesiuk [6].

Variant 3 corresponds to *water heating* with a heating coil placed in thermal insulation grooves, all covered *with a dry gypsum screed with* aluminium *lamellas* dissipating heat according to standard PN-EN 1264-2:2008 [4] or PN-EN 15377-1:2008 [5].

Variant 4 corresponds to *water heating* with a heating coil placed in thermal insulation grooves, all covered *with a "dry" cement screed, with steel lamellas* dissipating heat according to standard [4] or [5].

Variant 5 corresponds to *water heating* with a heating coil placed in the grooves of thermal insulation, all covered *with a layer of C2S1*-type adhesive, reinforced with glass fibre mesh *(without screed), without* heat-dissipating *lamellas*, based on the research of Karpiesiuk [6], Żukowski and Karpiesiuk [7], and Karpiesiuk and Chyży [8].

Variant 6 corresponds to *water heating* with a heating coil placed in thermal insulation grooves, all covered *with a layer of reactive adhesive (without screed), with aluminium lamellas* dissipating heat, based on the research of Karpiesiuk and Chyży [8].

Variant 7 corresponds to *electric heating* with heating cables laid *in "wet" concrete screed* as an accumulation system, using the second cheaper electricity tariff according to Electra's Vademecum [2].

Variant 8 corresponds to *electric heating* with *heating mats* placed in the C2S1 adhesive layer *on a "wet" concrete screed*, using electricity tariffs I and II according to Electra's Vademecum [2].

6.2 INPUT DATA SPECIFICATION

The necessary data were prepared in tables for each of the variants. At the end of all tables, there is information about the source literature on which the data were based. When the criteria are not measurable, the numerical data were adopted as comparative data for the considered variants. When controlling the heating, the influence of thermal inertia was taken into account (greater floor inertia makes it difficult to control underfloor heating). When assessing the ease of installation of the systems, the following were taken into consideration:

- execution of the installation using basic tools (without heavy equipment, e.g., concrete mixers and units for mixing concrete mortar)
- simplicity of installation following the assembly instructions
- labour intensity (e.g., weight of materia^ls used)

Variant 1 corresponds to traditional water heating with a heating coil placed in a "wet" concrete screed. Its basic parameters are given in Table 6.2.

TABLE 6.2
Specification of Underfloor Heating *with a Heating Coil in "Wet" Screed Concrete*

Specification

Heating Type	**Traditional – A Heating Coil Placed in Screed Concrete***
1. Net cost per 1 m^2 – material and labour + boiler (large areas) (1)	185 + 160 = PLN 345
2. Net cost per 1 m^2 – material and labour (e.g., renovation of small areas up to 20 m^2) (1)	PLN 310
3. The cost of operation and service for 15 years, taking into account the efficiency of the devices supplying the heating system (2)	PLN 58,050
4. Total investment cost per 100 m^2 of the heating system and operation over 15 years – the sum of 1. point, per 100 m^2 + 3. oint	PLN 92,550
5. Completion time: 100 m^2 with ceramic tile, including technological breaks (3)	49 days
6. Completion rate – 100 m^2/implementation time	2.04 m^2/day
7. Weight of 1 m^2 of the heating system without flooring	120 kg (2,000 kg/m^3 × 0.06 m)
8. Thermal inertia = ease of control – quick switching on and off (4), (6)	≈ 5 hours, for DT = 10K
9. Thermal efficiency at 35°C (5), (6)	67.27 W/m^2
10. Comfort of use = average temperature on the surface of the floor, at a supply temperature of 35°C (4), (6)	≈ 26.1°C
11. Versatility of carrying out renovation/modernisation without removing the concrete floor base – numerical rating (7)	1
12. Functional values = ease of control – numerical rating (8)	1
13. Ease of installation – numerical rating (8)	1
14. Manufacturer's warranty	10 years

*4.0 PLN/m^2 ≈ 1 \$/m^2 or 0.9 €/m^2.

Variant 2 corresponds to water heating with a heating coil placed in thermal insulation grooves, all covered with a "dry" gypsum screed, without heat-dissipating lamellas. Its basic parameters are given in Table 6.3.

Variant 3 corresponds to water heating with a heating coil placed in thermal insulation grooves, all covered with a "dry" gypsum screed, with aluminium lamellas dissipating heat. Its basic parameters are given in Table 6.4.

Variant 4 corresponds to water heating with a heating coil placed in thermal insulation grooves, all covered with a "dry" cement screed with steel lamellas dissipating heat. Its basic parameters are given in Table 6.5.

Variant 5 corresponds to water heating with a heating coil placed in thermal insulation grooves, all covered with a C2S1 adhesive-mesh layer (without screed), without heat-dissipating lamellas. Its basic parameters are given in Table 6.6.

Variant 6 corresponds to water heating with a heating coil placed in thermal insulation grooves, on aluminium lamellas dissipating heat. All of the materials are covered with a layer of reactive adhesive (without screed). Its basic parameters are given in Table 6.7.

TABLE 6.3
Specification of Underfloor Heating with Heating Coil Placed in Thermal Insulation Grooves, All Covered with *a "Dry" Gypsum Screed without Lamellas*

Specification

Heating Type	**Type B Heating – Heating Coil Placed in Thermal Insulation Grooves ***
1. Net cost per 1 m^2 – (material and labour + boiler (large areas)) (1)	230 + 160 = PLN 390
2. Net cost per 1 m^2 – material and labour (e.g., renovation of small areas up to 20 m^2) (1)	PLN 270
3. Operating cost (heating and service costs for 15 years) (2)	PLN 58,050
4. Total investment cost per 100 m^2 of the heating system, and operation over 15 years (the sum of 1. point per 100 m^2 + 3. point)	PLN 97,050
5. Completion time for 100 m^2 with ceramic tiles, including technological breaks (3)	21 days
6. Completion rate (100 m^2/implementation time)	4.76 m^2/day
7. Weight of 1 m^2 of the heating system without flooring	23.4 kg (Knauf Dual Floor)
8. Thermal inertia (ease of control – quick switching on and off) (4), (6)	≈ 2.5 hours, for DT = 10K
9. Thermal efficiency at 35°C (5), (6)	51.81 W/m^2
10. Comfort of use (average temperature on the surface of the floor at a supply temperature of 35°C) (4), (6)	≈ 25.55°C
11. Versatility, due to the ability to carry out renovation/modernisation without removing the concrete floor base – numerical rating (7)	3
12. Functional values (ease of control) – numerical rating (8)	3
13. Ease of installation (numerical rating) (8)	3
14. Manufacturer's warranty	10 years

*4.0 PLN/m^2 ≈ 1 \$/m^2 or 0.9 €/m^2.

TABLE 6.4
Specification of Underfloor Heating with a Heating Coil Placed in Thermal Insulation Grooves, All Covered with a *"Dry" Gypsum Screed with Aluminium Lamella*s

Specification

Heating Type	**Type B Heating – A Heating Coil Placed in Thermal Insulation Grooves ***
1. Net cost per 1 m^2 – material and labour + boiler (large areas) (1)	255 + 160 = PLN 415
2. Net cost per 1 m^2 – material and labour (e.g. renovation of small areas up to 20 m^2) (1)	PLN 310
3. Operating cost (heating and service costs for 15 years) (2)	PLN 58,050
4. Total investment cost per 100 m^2 of the heating system, and operation over 15 years – the sum of 1. point per 100 m^2 + 3. point	PLN 99,550
5. Completion time for 100 m^2 with ceramic tiles, including technological breaks (3)	21 days
6. Completion rate – 100 m^2/implementation time	4.76 m^2/day
7. Weight of 1 m^2 of the heating system, without flooring	23.4 kg (Knauf Dual Floor)
8. Thermal inertia (ease of control – quick switching on and off) (4), (6)	≈ 2 hours, for DT = 10K
9. Thermal efficiency at 35°**C** (5), (6)	71 W/m^2
10. Comfort of use (average temperature on the surface of the floor at a supply temperature of 35°C) (4), (6)	≈ 26.99°C
11. Versatility, due to the ability to carry out renovation/modernisation without removing the concrete floor base – numerical rating (7)	3
12. Functional values (ease of control) numerical rating (8)	3
13. Ease of installation – numerical rating (8)	3
14. Manufacturer's warranty	10 years

*4.0 PLN/m^2 ≈ 1 \$/m^2 or 0.9 €/m^2.

TABLE 6.5
Specification of Underfloor Heating with a Heating Coil in Thermal Insulation Grooves, All Covered *with a "Dry" Cement Screed with Galvanised Steel Lamella*s

Specification

Heating Type	**Type B Heating – A Heating Coil Placed in Thermal Insulation Grooves ***
1. Net cost per 1 m^2 – material and labour + boiler (large areas) (1)	430 + 160 = PLN 590
2. Net cost per 1 m^2 – material and labour (e.g., renovation of small areas up to 20 m^2) (1)	PLN 490
3. Operating cost (heating and service costs for 15 years) (2)	PLN 58,050

(*Continued*)

TABLE 6.5 (*Continued*)
Specification of Underfloor Heating with a Heating Coil in Thermal Insulation Grooves, All Covered *with a "Dry" Cement Screed with Galvanised Steel Lamellas*

Specification	
Heating Type	**Type B Heating – A Heating Coil Placed in Thermal Insulation Grooves ***
4. Total investment cost per 100 m² of the heating system, and operation over 15 years – the sum of 1. point per 100 m² and 3. point	PLN 117,050
5. Completion time 100 m² with ceramic tiles including technological breaks (3)	22 days
6. Completion rate – 100 m²/implementation time	4.55 m²/day
7. Weight of 1 m² of the heating system without flooring	37 kg (Knauf Aquapanel Floor, thickness 22 mm)
8. Thermal inertia (ease of control – quick switching on and off) (4), (6)	≈ 2 hours, for DT = 10K
9. Thermal efficiency at 35°C (5), (6)	87 W/m²
10. Comfort of use (average temperature on the surface of the floor at a supply temperature of 35°C) (4), (6)	≈ 27.24°C
11. Versatility, due to the ability to carry out renovation/ modernisation without removing the concrete floor base – numerical rating (7)	3
12. Functional values (ease of control) – numerical rating (8)	3
13. Ease of installation – numerical rating (8)	3
14. Manufacturer's warranty	10 years

*4.0 PLN/m² ≈ 1 $/m² or 0.9 €/m².

TABLE 6.6
Specification of Underfloor Heating with a Heating Coil in Thermal Insulation Grooves, All Covered with *C2S1 Adhesive, Reinforced with Mesh without Lamellas*

Specification	
Heating Type	**The Lightweight Heating System – A Heating Coil Placed in Thermal Insulation Grooves ***
1. Net cost per 1 m² – material and labour + boiler (large areas) (1)	190 + 160 = PLN 350
2. Net cost per 1 m² – material and labour (e.g., for renovation of small areas up to 20 m²) (1)	PLN 225
3. Operating cost (heating and service costs for 15 years) (2)	PLN 58,050

(*Continued*)

TABLE 6.6 (*Continued*)
Specification of Underfloor Heating with a Heating Coil in Thermal Insulation Grooves, All Covered with *C2S1 Adhesive, Reinforced with Mesh without Lamellas*

Specification

Heating Type	**The Lightweight Heating System – A Heating Coil Placed in Thermal Insulation Grooves ***
4. Total investment cost per 100 m^2 of the heating system and operation over 15 years – the sum of 1. point per 100 m^2 + 3. point	PLN 93,050
5. Completion time for 100 m^2 with ceramic tiles including technological breaks (3)	21 days
6. Completion rate – 100 m^2/implementation time	4.76 m^2/day
7. Weight of 1 m^2 of the heating system without flooring	6 kg (C2S1 adhesive)
8. Thermal inertia (ease of control – quick switching on and off) (4), (6)	≈ 1 hour, for DT = 10K
9. Thermal efficiency at 35°C (6)	75.3 W/m^2
10. Comfort of use (average temperature on the surface of the floor at a supply temperature of 35°C) (6)	≈ 27.0°C
11. Versatility due to the ability to carry out renovation/modernisation without removing the concrete floor base – numerical rating (7)	5
12. Functional values (ease of control – numerical rating) (8)	5
13. Ease of installation – numerical rating (8)	3
14. Manufacturer's warranty	10 years

*4.0 PLN/m^2 ≈ 1 \$/m^2 or 0.9 €/m^2.

TABLE 6.7
Specification of Underfloor Heating with a Heating Coil *Placed* in Thermal Insulation Grooves, All Covered by *Reactive Adhesive with Lamellas*

Specification

Heating Type	**Lightweight Heating System – A Heating Coil Placed in Thermal Insulation Grooves ***
1. Net cost per 1 m^2 – material and labour + boiler (large areas) (1)	240 + 160 = PLN 400
2. Net cost per 1 m^2 – material and labour (renovation of small areas up to 20 m^2) (1)	PLN 290
3. Operating cost (heating and service costs for 15 years) (2)	PLN 58,050
4. Total investment cost per 100 m^2 of the heating system plus operation over 15 years – the sum of 1. point per 100 m^2 + 3. point	PLN 98,050

(*Continued*)

TABLE 6.7 (*Continued*)
Specification of Underfloor Heating with a Heating Coil *Placed* in Thermal Insulation Grooves, All Covered by *Reactive Adhesive with Lamellas*

Specification

Heating Type	**Lightweight Heating System – A Heating Coil Placed in Thermal Insulation Grooves ***
5. Completion time for 100 m^2 with ceramic tiles including technological breaks (3)	14 days
6. Completion rate – 100 m^2/implementation time	7.14 m^2/day
7. Weight of 1 m^2 of the heating system without flooring	2.5 kg (Sika BondT8)
8. Thermal inertia = ease of control – quick switching on and off (4), (6)	≈ 1 hour, for DT = 10K
9. Thermal efficiency at 35°C (6)	129.5 W/m^2
10. User comfort (average temperature on the surface of the floor at a supply temperature of 35°C) (6)	≈ 33.4°C
11. Versatility, due to the ability to carry out renovation/ modernisation without removing the concrete floor base – numerical rating (7)	5
12. Functional values (ease of control) – numerical rating (8)	5
13. Ease of installation – numerical rating (8)	5
14. Manufacturer's warranty	10 years

*4.0 PLN/m^2 ≈ 1 \$/m^2 or 0.9 €/m^2.

Variant 7 corresponds to electric floor heating with heating cables laid in "wet" screed concrete, as an accumulation system using the second electricity, cheaper tariff. Its basic parameters are given in Table 6.8.

Variant 8 corresponds to the electric heating floor with heating mats placed in a C2S1 adhesive layer, on a "wet" concrete screed in electricity tariffs I and II. Its basic parameters are given in Table 6.9.

1. the average retail and wholesale prices of materials and labour using the same products and compatibility of assembly technologies given according to the article entitled "Lightweight Surface Radiator – Summary of Research in the Field of Sanitary Engineering", *Magazyn Instalatora*, 12 (244), December 2018 [9] adjusted according to market costs for March 2023. Water underfloor heating includes the connection of a gas boiler for PLN 16,000. net., where 4.0 PLN/m^2 ≈ 1 \$/m^2 or 0.9 €/m^2. The cost of small areas does not include the price of the boiler, assuming its existence, during renovation/modernisation. Electric heating is based on the manufacturer's calculations of heating cables and mats, following the technology according to market costs as of March 2023.
2. Electric heating systems. Vademecum Elektra, Table 2.1.6, p. 19. Issue 09/2021 [2]. The first and second electricity tariff applies to heating mats,

TABLE 6.8
Specification of Electric Underfloor Heating *(Heating Cable), Placed in "Wet" Screed Concrete*

Specification

Heating Type	**Electric Storage Heating – Heating Cable Place in Screed Concrete ***
1. Net cost per 1 m² – material and labour (1)	PLN 190
2. Net cost per 1 m² – material and labour (renovation of small areas up to 20 m²) (1)	PLN 225
3. Operating cost (heating and service costs for 15 years), where the total service cost is PLN 750 (2)	PLN 77,550
4. Total investment cost per 100 m² of the heating system, and operation over 15 years – the sum of 1. point per 100 m² + 3. point	PLN 96,550
5. Completion time 100 m² with ceramic tiles including technological breaks (3)	49 days
6. Completion rate – 100 m²/implementation time	2.04 m²/day
7. Weight of 1 m² of the heating system without floor	160 kg (2,000 kg/m³ × 0.08 m)
8. Thermal inertia (ease of control – quick switching on and off) (4), (6)	≈ 5 hours, for DT = 10K
9. Thermal efficiency at 35°C – (5), (6)	67.27 W/m² = variant I
10. User comfort = average temperature on the surface of the boards at a supply temperature of 35°C (4), (6)	≈ 26.1°C = variant I
11. Versatility, due to the ability to carry out renovation/modernisation without removing the concrete floor base – numerical rating (7)	1
12. Functional values (ease of control) – numerical rating (8)	1
13. Ease of installation – numerical rating (8)	1
14. Manufacturer's warranty	20 years

*4.0 PLN/m² ≈ 1 $/m² or 0.9 €/m².

TABLE 6.9
Specification of Electric Underfloor Heating (Heating Mats) Placed on Concrete Screed and Covered by C2S1 Adhesive

Specification

Heating Type	**Electric Heating Floor – Heating Mat Covered by an Adhesive Mortar ***
1. Net cost per 1 m² – material and labour + boiler (large areas) (1)	PLN 242
2. Net cost per 1 m² – material and labour (e.g., renovation of small areas up to 20 m²) (1)	PLN 300
3. Operating cost (heating and service costs for 15 years), where the total service cost is PLN 750 (2)	PLN 97,200

(Continued)

TABLE 6.9 (*Continued*)
Specification of Electric Underfloor Heating (Heating Mats) Placed on Concrete Screed and Covered by C2S1 Adhesive

Specification

Heating Type	**Electric Heating Floor – Heating Mat Covered by an Adhesive Mortar ***
4. Total investment cost per 100 m^2 of the heating system and operation over 15 years – the sum of 1. point per 100 m^2 + 3. point	PLN 121,400
5. Completion time – 100 m^2 with ceramic tiles, including technological breaks (3)	21 days
6. Completion rate – 100 m^2/implementation time	4.76 m^2/day
7. Weight of 1 m^2 of the heating system without floor	6 kg (C2S1 adhesive)
8. Thermal inertia (ease of control – quick switching on and off) (4), (6)	≈1.5 hours, for DT = 10K
9. Thermal efficiency at 35°C (5), (6)	75.3 W/m^2 = variant V
10. Comfort of use (average temperature on the surface of the boards at a supply temperature of 35°C) (4), (6)	≈ 27.0°C = variant V
11. Versatility, due to the ability to carry out renovation/modernisation without removing the concrete floor base – numerical rating (7)	5
12. Functional values (ease of control – numerical rating) (8)	5
13. Ease of installation – numerical rating (8)	3
14. Manufacturer's warranty	20 years

*4.0 PLN/m^2 ≈ 1 \$/m^2 or 0.9 €/m^2.

and the second electricity tariff applies to heating cables. The gas was applied to the boiler, and its cost was increased by 100% (annual heating costs) and by 50% (services) to adjust to costs in 2023. Example for gas fuel based on Table 2.1.6 – 1,710 + 100% = PLN **3,420** (annual heating costs at a rate of PLN 0.38/kWh gross) × 15 years = PLN 51,300 + PLN 4,500 + 50% = PLN 6,750 (service costs). Total 51,300 + 6,750 = PLN **58,050**. The annual costs of electricity heating are – 2,560 +100% = PLN **5,120** (tariff II - PLN 0.75/kWh gross) and 3,215 +100% = PLN **6,430** (tariffs I and II – average rate PLN 0.94/kWh gross), where 4.0 PLN/m^2 ≈ 1 \$/m^2 or 0.9 €/m^2.

The market costs assumed above are similar to, or higher than, those assumed in the updated Vademecum Elektra, Table 2.1.6, page 19, Edition 02/2024 [10]. In Table 2.1.6, the annual costs of heating with gas amount to PLN 2,682 (the assumed amount is PLN **3,420**), and the servicing costs for 15 years are equal to PLN 5,800 (the assumed amount is PLN 6,750). The total costs of gas heating and service in the 2024 catalogue amount to 40,230 + 5,800 = PLN 46,030, whereas the assumed total is PLN **58,050**. The annual costs of heating with electricity (tariff II) in the 2024 catalogue are PLN 5,345, and the assumed from market PLN **5,120** (they are similar), and in tariffs I and II they are PLN 6,305, and the assumed from market PLN **6,430** (they are similar), where PLN 4.0/m² ≈ 1 $/m² or 0.9 €/m².

Thus, it can be confirmed that the market costs for 2023 align with the calculations in the Elektra catalogue for 2024.

3. Assumptions for concrete drying – 28 days, laying heating and ceramic tiles per 100 m² – 21 days, with reactive adhesive, ceramic tiles that are immediately laid without a reinforcing primer – 14 days.
4. Thermal inertia, based on "How quickly does an underfloor heater cool down and heat up". Polish Installer. 2006.03. Authors: Prof. Ph.D. Eng. Marian Rosiński, MSc. Eng. Zenon Spik, Institute of Heating and Ventilation, Warsaw University of Technology [11].
5. Surface Heating/Cooling Boards. KAN-therm system. PL 03/2022. Table 5 from [12]. Underfloor heating efficiency – spout 4.5 cm above the pipe – diameter 16 × 2.0; t_{Fm} = 35°C - Profile, Tacker Rail; $R_{\lambda B}$ – thermal resistance of the floor covering [m²K/W] = 0.05, distance *T* between the coil 0.10 m.
6. Uniformity of temperature distribution, based on the engineering work entitled "Research on the Temperature Field during Surface Heating at the Białystok University of Technology".

   Contractor: Paweł Karpiesiuk. Supervisor: Prof. Ph.D. Eng. Mirosław Żukowski from [6].

   - Technical catalogue of Elektra Kardo products. January 2023, interpolating the heating coil spacing from 12.5 to 10 cm (system with lamellas) by increasing the thermal power and floor temperature by 10%, $R_{\lambda B}$ – thermal resistance of the floor covering [m²K/W] = 0.05 from [1].
   - Żukowski M., Karpiesiuk P., Test results of type B surface heating made using dry technology, *Instal* 7/8 (2015), pp. 33–37 of [13].
   - Żukowski M., Karpiesiuk P., Test results of a surface radiator with a very low height, *Instal* 10 (2015), pp. 38–41 of [7].
7. Numerical assessment of the universality of execution: 1 – impossible, 3 – sometimes, and 5 – possible.
8. Numerical rating for ease of control and assembly: 1 – difficult, 3 – medium, and 5 – easy.

6.3 OBJECTIVE FUNCTIONS AND INPUT CRITERIA

The aim of the MCDA is to select from a set of developed variants the most advantageous underfloor heating system for use in a single-family house, in terms of the investor's requirements, after adopting the assumed criteria, checking the data by a group of experts and using experimental research. MCDA was performed using the following literature [14–17]. It should be mentioned that, in the literature, there are also other criteria for assessing the selection of the best variant, such as multi-attribute utility theory (e.g., AHP – example in [18], ANP, MUZ, and DEMATEL methods), relational methods (e.g., ELECTRE, DELPHI, and PROMETHEE methods), and many others (e.g., geometric methods), widely described in the book [19].

When describing the object of assessment, one should answer the question: what features or criteria of this object are important? These may be physical, chemical, technical, economic, aesthetic, functional, etc.

In the evaluation process, both non-measurable criteria (which cannot be directly expressed numerically) and measurable criteria (expressed in numerical quantities) can be used. Non-measurable criteria are quantified and ultimately expressed in numerical values, which allows for their comparison and assessment. For this purpose, an arbitrarily adopted grading scale is introduced. It is defined and given under the tables describing the eight analysed variants of underfloor heating.

In the multi-criteria evaluation process, the nature of the adopted K_j criteria is determined first. A criterion is called a stimulant if its higher value results in a more favourable overall assessment of the quality of the object. A criterion whose higher value worsens the overall assessment of the quality of an object is called a destimulant. Taking into account the above definitions, a Preliminary Set of Criteria {PSC} was created for the considered object of analysis. Each criterion specifies its information-carrying capacity, interdependencies between the criteria, and their nature.

6.3.1 The List and Description of Criteria

K1 – investment cost for 1 m^2 of underfloor heating system

- allows for a comparative assessment of solutions in terms of the necessary financial outlays for the purchase of materials for the heating system, as well as its implementation;
- depends on the materials used in the heating system, resulting from the installation technology and assuming the same thermal insulation and power of heating devices;
- measurable criterion;
- destimulant.

K2 – investment cost in small areas

- allows one to compare the profitability of implementing a heating system in specific rooms with small investments, during renovation/modernisation (e.g., bathroom + kitchen up to 20 m^2);
- measurable criterion;
- destimulant.

K3 – the cost of operating the underfloor heating system over 15 years of use, taking into account the efficiency of the devices supplying the system heating

- allows one to assess the amount of expenditure needed to maintain a comfortable temperature with various heating systems and media (water, electric). When using heating cables, storage heating is included in the second electricity tariff, heating mats in the first and second electricity tariffs;
- measurable criterion;
- destimulant.

K4 – total investment cost per 100 m^2, and operation over 15 years;

- allows one to assess investment and operating expenses over 15 years;
- measurable criterion;
- destimulant.

K5 – construction time of 100 m^2 of the heating system, including the ceramic tile floor, taking into account technological breaks

- allows for the assessment of technical and organisational solutions to the problem of speed of installation of the heating system with technological breaks (e.g., inability to lay the floor due to the concrete drying process being slow);
- measurable criterion;
- destimulant.

K6 – the pace of installation of the heating system

- the speed of the installation process contributes to an earlier commissioning of the investment. This reduces concerns about timely completion of work and eliminates financial penalties due to delays. They were calculated according to 100 m^2/K5;
- measurable criterion calculated as 100 m^2/execution time;
- stimulant.

K7 – the weight of the heating system per 1 m^2

- the greater weight of the system results in a higher cost of construction (thicker ceiling or a higher degree of reinforcement) and also increases the inertia of the heating, which makes it more difficult to control. The weight of the screed or adhesive is given depending on the variant;
- measurable criterion;
- destimulant.

K8 – thermal inertia of the heating system, which translates into quality control

- avoids overheating and underheating, thanks to a quick response to turning off the heating system in winter or turning on cooling in summer. Low thermal inertia was taken into consideration, allowing for maintaining a comfortable temperature inside, and at the same time providing measurable economic benefits during operation;
- measurable criterion;
- destimulant.

K9 – thermal efficiency at a low supply temperature of 35°C

- allows one to assess the heat flux density (heat efficiency at low temperatures of the heating medium) important in RES;
- measurable criterion;
- stimulant.

K10 – the comfort of use, depending on the average temperature of the floor at a supply temperature of 35°C

- the comfort is assessed based on the floor temperature relative to the supply temperature;
- measurable criterion;
- stimulant.

K11 – the universality of installation method during renovation or modernisation, without removal of the concrete base

- allows the system to be easily implemented in both new and modernised buildings, taking into account the level of the heating floor;
- not a measurable criterion, a three-point assessment scale was adopted;
- stimulant.

K12 – functional values (ease of control)

- ease of control, depending on the thermal inertia of the floor;
- not a measurable criterion, a 3-point rating scale was adopted;
- stimulant.

K13 – ease of installation

- ease of installation affects the speed of investment implementation;
- not a measurable criterion, a 3-point grading scale was adopted;
- stimulant.

K14 – manufacturer's warranty

- the warranty proves the quality of the materials used, including their durability and the provision of the manufacturer/importer's service in the event of possible failures (faults), which is an important factor in the comfort of use;
- measurable criterion;
- stimulant.

The selected criteria served as the basis for further analysis.

6.4 THE SET OF CRITERIA WITH THEIR ANALYSIS AND VERIFICATION

In decision-making situations, when the subject of the assessment is described by more than a few criteria, the analysis of the initially adopted set is difficult, especially in terms of the mutual relations between the criteria. In this case, it is recommended to use a criteria matrix and graph.

The construction of the criteria matrix $G_{(n+1,\, n+1)}$, where n is the number of criteria, is performed by analysing each pair of criteria K_i, K_j $(i \neq j)$ in terms of the information they provide in the assessed solution. If this information overlaps even partially, the direction of this relationship is examined. When the criterion K_j contains (i.e., carries) information about K_i, we encode the information by assigning the value 0 to the element g(i, j) and the value 1 to the element g(j, i) of the matrix being built (i – row number, j – column number). If the information K_i and K_j are independent, the value 0 is assigned to the elements g(i, j) and g(j, i). In this way, the matrix presented in Table 6.10 was created.

6.4.1 Explanations for Selected Criteria

Criterion K4, the total cost of investment and operation, contains (carries) part of the information about the criterion K1 (investment cost) and K3 (operating cost), then the element g(4, 1) = 1, the element g(1, 4) = 0, and g(4, 3) = 1, element g(3, 4) = 0.

Criterion K5, the time of construction of the heating system, taking into consideration technological breaks, contains (carries) part of the information about

TABLE 6.10
Matrix $G_{15.15}$ Relationships between the Criteria K_1, K_2, … K_{14}

No.	1	2	3	4	5	6	7	8	9	10	11	12	13	14	15	b_i
1	X	0	0	0	0	0	0	0	0	0	0	0	0	0	0	0
2	0	X	0	0	0	0	0	0	0	0	0	0	0	0	0	0
3	0	0	X	0	0	0	0	0	0	0	0	0	0	0	0	0
4	1	0	1	X	0	0	0	0	0	0	0	0	0	0	0	2
5	0	0	0	0	X	1	0	0	0	0	0	0	0	0	0	1
6	0	0	0	0	0	X	0	0	0	0	0	0	0	0	0	0
7	0	0	0	0	0	0	X	1	0	0	0	0	0	0	0	1
8	0	0	0	0	0	0	0	X	0	0	0	1	0	0	0	1
9	0	0	0	0	0	0	0	0	X	0	0	0	0	0	0	0
10	0	0	0	0	0	0	0	0	0	X	0	0	0	0	0	0
11	0	0	0	0	0	0	0	0	0	0	X	0	0	0	0	0
12	0	0	0	0	0	0	0	0	0	1	0	X	0	0	0	1
13	0	0	0	0	1	0	0	0	0	0	0	0	X	0	0	1
14	0	0	0	0	0	0	0	0	0	0	0	0	0	X	0	0
15	1	1	1	1	1	1	1	1	1	1	1	1	1	1	X	15
b_j	2	1	2	1	2	2	1	2	1	2	1	2	1	1	0	X

criterion K6. The pace of implementation depends on the duration of the investment, then the element g(5, 6) = 1 and the element g(6, 5) = 0.

Criterion K7, the weight of the heating system, contains (carries) part of the information about criterion K8 – thermal inertia (depending on the mass of the material used), then the element g(7, 8) = 1 and the element g(8, 7) = 0.

Criterion K8, thermal inertia, carries information about K12 (ease of control, when the inertia is large, heating control is more difficult), then the element g(8, 12) = 1 and the element g(12, 8) = 0.

Criterion K12, ease of control, carries information about K10 (the comfort of use-when control is easy, comfort increases), then the element g(12, 10) = 1 and the element g(10, 12) = 0.

Criterion K13, ease of assembly, provides information about K5 – execution time (when we have a simple assembly, its time is shorter), then the element g(13, 5) = 1 and the element g(5, 13) = 0.

An additional element was also introduced to the set of criteria, the so-called global criterion K_{n+1}, assigning the value of 1 to the elements of $G_{(n+1,k)}$ and the value of 0 to the elements of $G_{(k,n+1)}$. The coding of the dependencies between the criteria in this way is presented in the form of the matrix G{15,15} in Table 3.9. The dimension of the matrix results from the number of criteria – 14 partial criteria and 1 global criterion.

Based on the developed matrix, a criteria graph was built. First, the number of arcs entering each criterion (vertex) of the graph was calculated.

This value was calculated according to the formula:

$$b_j = \sum_{i=1}^{n+1} g_{ij} \tag{6.1}$$

where

b_j – number of arcs entering the j-th vertex,

g_{ij} – elements of matrix G.

Then, the number of arcs coming from the i-th vertex was calculated:

$$b_i = \sum_{j-1}^{n+1} g_{ij} \tag{6.2}$$

where

b_i – number of arcs coming from the i-th vertex,

g_{ij} – elements of matrix G.

The values *of* b_j and b_i are given in the matrix $G_{15,15}$ as the 16th row and 16th column, respectively. Next, the layered ordering of the graph was performed according to the following algorithm. Layer 0 of the graph is created by the global criterion K_G, which is connected by arcs directed to each of the other criteria.

Layer 1 of the graph consists of criteria having one arc, i.e., criteria for which $b_j = 1$. These are the criteria: K2, K4, K7, K9, K11, K13, and K14, which are connected by arcs to the global criterion K_G.

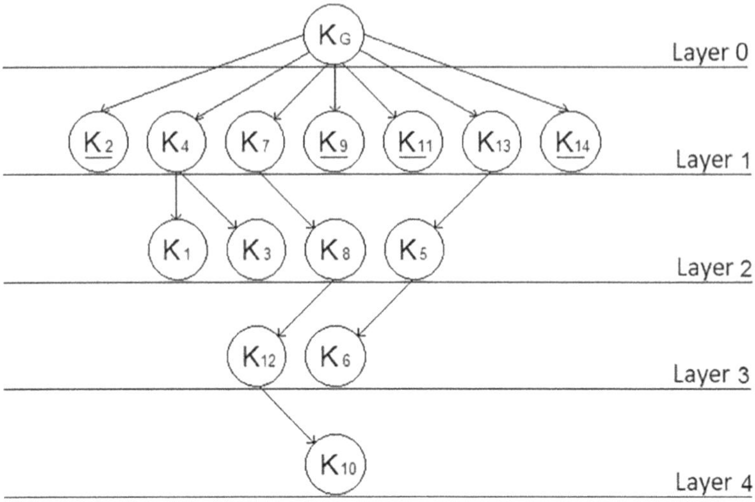

FIGURE 6.1 Layer-ordered criteria graph according to matrix G {15,15}.

To arrange subsequent layers, we select criteria that include arcs from some vertices, but only from the layers located above. As can be seen in the matrix $G_{15.15}$, layer 2 of the graph consists of criteria K1, K3, K5, and K8. Layer 3 of the graph consists of criteria K6 and K12. The last 4th layer of the graph is created by the K10 criterion. The vertices of the last layer have no outgoing arcs. This means the graph is complete. A criteria graph illustrating the relationship between the criteria is shown in Figure 6.1.

Layering the graph makes it easier to create the final set of criteria {FSC} so that the set is representative and contains as few elements as possible.

Based on the developed matrices and criteria graph, those that allow for the creation of a sufficiently complete and representative description of the assessed decision variants were selected.

{FSC} include all criteria that are connected only to the vertex of the graph, i.e., the global criterion K_G. These are

K2 – investment cost in small areas [PLN]
K9 – thermal efficiency
K11 – universality of execution during renovation or modernisation
K14 – manufacturer's warranty

For {FSC}, in addition to those mentioned above, the criteria included were those found in the layers with the largest number of vertices. These are the criteria from layer 2:

K1 – investment cost per 1 m^2 [PLN]
K3 – the cost of operating system over 15 years
K5 – construction time of 100 m^2 of the heating system
K8 – thermal inertia of the heating system

Criteria selected in this way help avoid multiple considerations of the same information in the assessment and provide as few elements as possible.

6.5 THE WEIGHTS OF CRITERIA WITH THEIR ANALYSIS AND VERIFICATION

Partial criteria measures for individual variants of the designed solutions were determined based on data from Tables 6.2 to 6.9, listing the specifications of each of the eight variants considered, adopted according to those described under the documentation tables.

The measures were determined based on an arbitrarily adopted rating scale described in Tables 6.2–6.9 for non-measurable criteria. The values of the measures and the characteristics of the criteria are presented in Table 6.11.

In multi-criteria analysis, it should be remembered that the adopted criteria often have different importance. This fact is taken into account in the evaluation algorithm. For this purpose, weights ν_i are introduced to correct the values of the criteria according to the decision-maker's preferences and, if necessary, after consultation with a group of experts or by analysing pairs.

The process of determining the weights, supported, for example, by pair analysis, involves building a square matrix A with dimensions n × n, where n is the number of criteria. The elements a_{ij} of this matrix are assigned a value depending on the preferences of the criteria k_i and k_j specified by the decision-maker:

$a_{ij} = 1$ when k_i is more important than k_j

$a_{ij} = 0$ in other cases

By summing the elements of matrix A along the row, we obtain a certain weight indicator of the i-th criterion, which can be treated directly as the weight of the i-th criterion or transformed so that the value of the weight indicator belongs to the desired range, e.g., (0, 1).

TABLE 6.11
Criteria Measures for the Considered Variants

Variants \ Criterion[a]	K1, d	K2, d	K3, d	K5, d	K8, d	K9, s	K11, s	K14, s
Variant 1	345	310	58,050	49	5	67.27	1	10
Variant 2	390	270	58,050	21	2.5	51.81	3	10
Variant 3	415	310	58,050	21	2	71	3	10
Variant 4	590	490	58,050	22	2	87	3	10
Variant 5	350	225	58,050	21	1	75.3	5	10
Variant 6	400	290	58,050	14	1	129.5	5	10
Variant 7	190	225	77,550	49	5	67/27	1	20
Variant 8	242	300	97,200	21	1.5	75.3	5	20

[a] Designations for the K criteria, where d – destimulant and s – stimulant.

The sum of the criterion weights in a given study was assumed to be equal to 1. It was decided to distinguish five sets of weights, characterising the selected criteria priorities in different ways. Sets III and V have the same highest priorities of importance – investment and operation, but in set III, the remaining criteria are included but less important, and in set V, completely different criteria are omitted. For this reason, the general analysis of all sets of weights was from I to IV, and the V set was treated as confirming the correctness of MCDA calculations. Paired importance analysis was used in sets of weights I, II, and IV, as shown in Tables 6.12–6.14.

The weights for the K1 and K2 criteria concern investment costs, divided into those relating to the entire building – K1 and small areas (under renovation) – K2. When we analyse a large investment, the weights K2 equal 0, and when renovating or modernising the scales, K1 = 0.

TABLE 6.12
Matrix of Weights I, the Importance of Implementation Time, and Universality of Heating Implementation in Pair Analysis

X	K1	K2	K3	K5	K8	K9	K11	K14	a_i	$a_i/\Sigma a_i$
K1	X	1	0	0	0	0	0	1	2	0.091
K2	0	X	0	0	0	0	0	0	0	0
K3	0	1	X	0	0	0	0	1	2	0.091
K5	1	1	1	X	1	1	1	1	7	0.319
K8	0	1	0	0	X	0	0	1	2	0.091
K9	0	1	0	0	0	X	0	1	2	0.091
K11	1	1	1	0	1	1	X	1	6	0.273
K14	0	1	0	0	0	0	0	X	1	0.044
Σa_i									22	X

TABLE 6.13
Matrix of Weights II, Importance of Investment Price and Thermal Efficiency in Pair Analysis

X	K1	K2	K3	K5	K8	K9	K11	K14	a_i	$a_i/\Sigma a_i$
K1	X	0	1	1	1	1	1	1	6	0.3
K2	0	X	0	0	0	0	0	0	0	0
K3	0	0	X	0	0	0	0	1	1	0.05
K5	0	0	1	X	0	0	0	0	1	0.05
K8	0	0	1	1	X	0	0	1	3	0.15
K9	0	0	1	1	1	X	1	1	5	0.25
K11	0	0	1	1	0	0	X	0	2	0.1
K14	0	0	1	0	0	0	1	X	2	0.1
Σa_i									20	X

TABLE 6.14
Matrix of Weights IV, When Heating a Small Area (Planned Renovation) in Pair Analysis

X	K1	K2	K3	K5	K8	K9	K11	K14	a_i	$a_i/\Sigma a_i$
K1	X	0	0	0	0	0	0	0	0	0
K2	0	X	1	0	1	1	0	1	4	0.222
K3	0	0	X	0	0	0	0	1	1	0.056
K5	0	1	1	X	1	1	0	1	5	0.277
K8	0	0	0	0	X	0	0	1	1	0.056
K9	0	0	0	0	0	X	0	1	1	0.056
K11	0	1	1	0	1	1	X	1	5	0.277
K14	0	0	1	0	0	0	0	X	1	0.056
Σa_i									18	X

TABLE 6.15
Sets of Criteria Weights

Variants	Set of Weights I	Set of Weights II	Set of Weights III	Set of Weights IV	Set of Weights V
ν_1 – weight of the K1 criterion	0.091	0.3	0.25	0	0.5
ν_2 – weight of the K2 criterion	0	0	0	0.222	0
ν_3 – weight of the K3 criterion	0.091	0.05	0.25	0.056	0.5
ν_4 – weight of the K5 criterion	0.319	0.05	0.1	0.277	0
ν_5 – weight of the K8 criterion	0.091	0.15	0.1	0.056	0
ν_6 – weight of the K9 criterion	0.091	0.25	0.1	0.056	0
ν_7 – weight of the K11 criterion	0.273	0.1	0.1	0.277	0
ν_8 – weight of the K14 criterion	0.044	0.1	0.1	0.056	0

The following priorities were selected per the number of weights:

- Set I of criteria weights – the importance of implementation time and universality of heating was assumed;
- Set II of criteria weights – the importance of investment cost and thermal efficiency was assumed;
- Set III of criteria weights – the importance of investment and operating costs was assumed;
- Set IV of criteria weights – heating in a small area was assumed (planned renovation);
- Set V of criteria weights – only the investment and operating costs were assumed to be valid as in set III, but with the omission of other criteria.

The method of using the five sets of scales is described in Table 6.15.

6.6 SELECTION OF THE MCDA METHODOLOGY ADEQUATE TO THE SET OF CRITERIA

Among the MCDA methods, there is a group of algorithms that construct a scalar whose numerical value serves as a synthetic evaluation indicator. These are the most commonly used mathematical methods for solving multi-criteria problems. These methods are the simplest and fully meet the requirements for MCDA.

Mathematical methods are a quick and effective tool for optimising solutions. It is recommended to conduct multi-criteria comparative analyses based on the summative indicator, adjusted summative indicator, additive indicator, adjusted additive indicator, or weighted arithmetic mean. Synthetic evaluation indices are calculated using the following formulas:

Summative indicator:

$$J_i = \sum_{j=1}^{m} z_{ij} \tag{6.3}$$

Adjusted summative indicator:

$$J_i = \sum_{j=1}^{m} (z_{ij} \cdot v_j) \tag{6.4}$$

Additive indicator:

$$J_i = \frac{1}{m} \sum_{j=1}^{m} z_{ij} \tag{6.5}$$

Adjusted additive indicator:

$$J_i = \frac{1}{m} \sum_{j=1}^{m} (z_{ij} \cdot v_j) \tag{6.6}$$

Weighted arithmetic mean:

$$J_i = \frac{\sum_{j=1}^{m} (z_{ij} \cdot v_j)}{\sum_{j=1}^{m} v_j)} \tag{6.7}$$

where

z_{ij} – coded measure of the j-th criterion for the i-th variant;

ν_j – the weight of the j-th criterion;

m – number of criteria.

The adjusted summary index was chosen to carry out the analysis in this work, which allows one to take into account the weights of the various criteria and more accurately select the most advantageous ones. It does not require that the coded criterion measures have values greater than zero. The considered criteria are designated values (e.g., for criterion K1 – investment cost and K5 – order completion time); partial measures are expressed in [PLN] and [days], respectively. Therefore, it is not possible to perform any mathematical operations without first coding the values from the data matrix.

6.7 PRESENTATION OF THE COMPUTATIONAL ALGORITHM

Coding the criteria values involves reducing the nominated values to the non-nominated ones, which makes it possible to conduct a comparative analysis and evaluate the solution sets. The criteria of stimulants and destimulants are considered separately.

Normalisation was chosen as the coding method in this work, assuming maximisation, i.e., reducing the criteria values from decreasing to increasing ones. Normalisation brings the nominal values to the range [0,1]. It also ensures that the coded criteria measures have values greater than zero, which expands the possibilities of using other synthetic indicators for analysis.

The norming procedure aims to replace the partial measure X_{ij} with its ratio to the maximum value of the measure according to the K_j criterion for all variants.

- for stimulants

  measure of the i-th variant according to the j-th criterion:

$$z_{ij} = \frac{x_{ij}}{x_{jmax}} \tag{6.8}$$

- for destimulants

  measure of the i-th variant according to the j-th criterion:

$$z_{ij} = \frac{x'_{ij}}{x'_{j,max}} \tag{6.9}$$

where reducing decreasing values to increasing ones (maximisation):

$$x'_{ij} = \frac{1}{x_{ij}} \tag{6.10}$$

$x'_{j,max}$ – maximum value of the measure of the j-th criterion after maximisation.

For destimulants, we reduce the values of decreasing features given in Table 6.16 to increasing ones using formula (6.10) and find the maximum values of each criterion max K_j.

Using the above formulas (6.8) and (6.9) and data from Table 6.16, the coded values of all criteria were calculated and are given in Table 6.17.

TABLE 6.16
Corrected Nominal Values of the Criteria for the Considered Variants

Criterion / Variants	K1	K2	K3	K5	K8	K9	K11	K14
	Destimulant					Stimulant		
Variant 1	0.0028986	0.003226	1.72E-05	0.02041	0.2	67.27	1	10
Variant 2	0.0025641	0.003704	1.72E-05	0.04762	0.4	51.81	3	10
Variant 3	0.0024096	0.003226	1.72E-05	0.04762	0.5	71	3	10
Variant 4	0.0016949	0.002041	1.72E-05	0.04546	0.5	87	3	10
Variant 5	0.0028571	0.004444	1.72E-05	0.04762	1	75.3	5	10
Variant 6	0.0025	0.003448	1.72E-05	0.07143	1	129.5	5	10
Variant 7	0.0052632	0.004444	1.29E-05	0.02041	0.2	67.27	1	20
Variant 8	0.0041322	0.003333	1.03E-05	0.04762	0.67	75.3	5	20
max K_j	**0.0052632**	**0.004444**	**1.72E-05**	**0.07143**	**1**	**129.5**	**5**	**20**

TABLE 6.17
Coded Values of Criteria Measures for Individual Variants

Criterion / Variants	K1	K2	K3	K5	K8	K9	K11	K14
	Destimulant					Stimulant		
Variant 1	0.550725	0.725806	1	0.28571	0.2	0.519459	0.2	0.5
Variant 2	0.487179	0.833333	1	0.666662	0.4	0.400077	0.6	0.5
Variant 3	0.457831	0.725806	1	0.666662	0.5	0.548263	0.6	0.5
Variant 4	0.322034	0.459184	1	0.636366	0.5	0.671815	0.6	0.5
Variant 5	0.542857	1	1	0.666662	1	0.581467	1	0.5
Variant 6	0.475	0.775862	1	1	1	1	1	0.5
Variant 7	1	1	0.748549	0.28571	0.2	0.519459	0.2	1
Variant 8	0.785124	0.75	0.597222	0.666662	0.67	0.581467	1	1

6.8 DISCUSSION AND INTERPRETATION OF THE ANALYSIS RESULTS

Using Table 6.17, and taking into account the weights of the criteria given in Section 6.5, the calculation of synthetic indicators was carried out according to formula (3.4) for the eight variants considered. The calculation was not made by multiplying the lines from Table 6.17 with weight column. The sum of the products gives the value of the summary index, adjusted for the appropriate variant (row) and the selected set of weights. The summary index J_i is a numerical indicator of the quality of the variant. The higher it is, the better the variant. Paired importance analysis was used in sets of weights I, II, and IV. In set III, the investment and operating costs were assumed to be significant at the level of 0.25, the remaining criteria were rated equally as 0.1, and in set V, the investment and operating costs were assumed to be 0.5, and the remaining criteria were omitted from the calculations.

TABLE 6.18
The Results of a Multi-Criteria Decision Analysis

Variants	Set of Criteria Weights I	Set of Criteria Weights II	Set of Criteria Weights III	Set of Criteria Weights IV	Sum of Weights I–IV	Set of Criteria Weights V	The Place
W1 (traditional heating)	0.374	0.459	0.558	0.420	1.811	0.775	
W2 (dry screed without lamellas)	0.607	0.500	0.629	0.665	2.401	0.744	
W3 (dry screed with lamellas)	0.627	0.543	0.646	0.655	2.471	0.729	
W4 (dry screed with steel lamellas)	0.616	0.531	0.621	0.594	2.362	0.661	
W5 (without screed and lamellas)	0.792	0.692	0.761	0.856	3.101	0.771	Second place
W6 (without screed with lamellas)	**0.930**	**0.793**	**0.819**	**0.922**	**3.464**	0.738	**First place**
W7 (electric storage)	0.414	0.632	0.658	0.495	2.199	0.874	
W8 (electric heating mats)	0.769	0.745	0.737	0.788	3.039	0.691	Third place
First place **W6**	**W6**	**W6**	**W6**	**W6**		*W7*	
Second place **W5**	W5	W8	W5	W5		W1	
Third place **W8**	W8	W5	W8	W8		W5	

The results of the multi-criteria comparative analysis are presented in Table 6.18.

After calculating the adjusted summary index J_i and assigning weights to the criteria in **set I**, taking into account the importance of implementation time due to financial penalties for untimely execution of the investment and taking into account the versatility of heating in various conditions, the most advantageous variant turned out to be **variant 6,** with the maximum value of $J_6 = 0.930$, the second most advantageous was variant 5, the third was variant 8, and the rest according to the value of the J_i and were listed in the row: W3 > W4 > W2 > W7 > W1.

An important factor when implementing heating systems is the investment cost of ensuring high thermal efficiency of heating. Here, when assessing the criteria weights in **set II,** the most important features are criterion 1 – investment cost, criterion 9 – thermal efficiency, criterion 8 – influencing the efficiency, and thermal inertia, for which higher weights were assigned than the others according to couple analysis The criteria weights were adopted for $v_1 = 0.30$, $v_9 = 0.25$, $v_8 = 0.15$, v_{11} and $v_{14} = 0.10$, respectively. The remaining two criteria were given equal weight, 0.05

each, as the least important when assessing the costs and efficiency of the system. After calculations, the most advantageous variant turned out to be variant 6 again, with the maximum value of $J_8 = 0.793$. In second place was variant 8, and in third place was variant 5. The remaining variants were in the row: W7 > W3 > W4 > W2 > W1.

Looking into the future, it is worth analysing not only the investment costs but also future expenses related to the operation of the building's heating. Therefore, criterion K1 was extended to include operating costs included in criterion K3, adopting **a set of weights III**, where higher weights of 0.25 were given to investment costs K1 and operating costs K3, and in the remaining criteria, their equal importance of 0.1 was assumed. After calculation, the most advantageous variant was once again variant 6, with the maximum value of $J_6 = 0.819$. The second place was taken by variant 5, and the third place by variant 8. The remaining variants were in the row: W7 > W3 > W2 > W4 > W1.

During construction, renovation, or modernisation, work related to underfloor heating is often carried out as well. It usually concerns smaller areas, when it is desired to improve the comfort conditions in the bathroom, kitchen, or vestibule, by installing a surface radiator. For this reason, **set IV of weights was adopted for the analysis**. In this case, the most important features are time and the universality of implementation. The investor expects the contractor to complete the work as quickly as possible, with the least amount of demolition work. The cost of renovation/modernisation is also important. After the pair analysis, the highest weights were adopted for the criteria K5 and K11 – appropriate time and the universality of implementation at the level of 0.277, and K2 – the cost of investment in small areas equal to 0.222. The remaining criteria were assessed equally with the same weight of 0.056. And in this case, once again, the best variant is variant 6, and then variants 5 and 8. The remaining variants were in the row: W2 > W3 > W4 > W7 > W1.

Additionally, the fifth set of criteria weights was taken into account, where the investment and operating costs were analysed as in the case of the third set, but at a weight level of 0.5. Here, the remaining criteria were completely omitted from the calculations. In this selection of scales, the first place was taken by variant 7 (electric storage heating), the second by variant 1, the so-called traditional – most often used in investment projects, and the third by variant 5. The remaining variants were in the row: W2 > W6 > W3 > W8 > W4. These results confirmed the correctness of the calculations, because the variants that took the top positions had the lowest overall investment and operating costs.

After taking into account the full universal MCDA evaluation list, it turned out that *option 6, i.e., water heating without screed with lamellas, turned out to be the most advantageous solution* in all four sets of weights; therefore, it was selected as the final one.

To summarise the MCDA results:

1. The MCDA is a simple and effective tool for organising solutions to these problems and selecting the best form of underfloor heating systems in residential buildings during their construction or renovation.
2. Four sets of criteria weights were adopted to enable the identification of universal preferences, namely, the cost of investment and operation, efficiency

of the heating system, time, and versatility of implementation in the conditions of a new building and during renovation on small areas to improve the comfort of use. The fifth set of criteria weights was adopted as a test of the correctness of the calculations, where only two criteria were taken into consideration.

3. The criteria weights were selected by an analysis of the pair's importance, and, in the set regarding investment and operating costs, the importance of these criteria was taken into consideration.
4. In adjusted summary indices, J_i individual sets of criteria weights were calculated and then summed to select the best universal variant for all criteria through pair analysis and in two subsequent places, II–III. Here are the individual results:

   First place – variant 6 – (water heating floor system using reactive adhesive without screed, with aluminium lamellas). $\Sigma J_6 \approx 3.46$;

   Second place – variant 5 – (water heating floor system using C2S1 adhesive without screed, without lamellas). $\Sigma J_5 \approx 3.10$;

   Third place – variant 8 – (electric heating floor system with heating mats using C2S1 adhesive on the screed). $\Sigma J_8 \approx 3.04$;
5. Additionally, individual places were also determined depending on the sum of *the highest* J scores, and related to the priorities of each set with criteria weights I–IV (vertical):

   First place – variant 6 – (set of weights I, II, III, IV). $\Sigma J_6 \approx 3.46$;
   Second place – variant 5 – (set of weights I, III, IV). $\Sigma J_5 \approx 2.41$;
   Third place – variant 8 – (set of weights I, III, IV). $\Sigma J_8 \approx 2.29$.
6. It turned out that the traditional, most commonly used water underfloor heating system (**variant 1**) is not a beneficial solution. After calculations were made in which a set of criteria weights regarding the investment and operating costs was adopted equal to 0.5 of these criteria, and the remaining ones were not taken into account, this type of heating took the **second place. The first place** went to the storage electric heating system using the cheap second electricity tariff (**variant 7**), which confirms the quality of MCDA calculations, as both systems are the cheapest in operation and investment.
7. When an additional set of criteria weights was adopted (in addition to the adopted analysis) for each criterion at the same level of 0.143, **the best** solutions turned out to be **variants 6, 8, and 5,** i.e., those in which the heating system is located just under the floor with $J_6 = 0.854$, $J_8 = 0.758$, and $J_5 = 0.757$.
8. The analysis of all variants confirmed the choice of **variant 6,** i.e., **water heating floor system with reactive adhesive without screeds and aluminium lamellas,** as the most advantageous one. It was selected as the final choice of the underfloor heating system in a 100-m^2 residential house, and when it is necessary to renovate or modernise smaller rooms in the facility, taking into account the universality of the selection of weights based on the pair analysis and the MCDA calculation procedure.

Variant 5 is also a favourable solution, i.e., **water heating floor system using C2S1 glue without screed and without lamellas**, which was ranked the **second in MCDA**.

ABBREVIATIONS

AHP	analytic hierarchy process
ANP	analytic network process
C2S1	deformable cement adhesive mortar
C2S2	deformable cement adhesive mortar
DELPHI	name of modern expert decision-making
DEMATEL	decision-making trial and evaluation laboratory
ELECTRE	elimination and choice translating reality
FSC	final set of criteria
MCDA	multi-criteria decision analysis
MUZ	Metoda Uogólnionej Zmiennej (in Polish)
PE-RT/AL/PE-RT	multilayer polyethylene pipe with an aluminium insert
PLN	Polish Zloty
PROMETHEE	Preference Ranking Organisation Method for Enrichment Evaluation
PSC	preliminary set of criteria
XPS	extruded polystyrene
$	American Dollar
€	European Euro

REFERENCES

1. Elektra Kardo technical catalogue, Poland, January 2023.
2. Elektra Vademecum, Poland, edition 09/2021.
3. Thullner K., *Low-energy buildings in Europe: standards, criteria and consequences: a study of nine European countries*. Avdelningen for installationsteknik. Institutionen för bygg- och miljöteknologi. Lunds tekniska högskola. Lunds Universitet, Lund, 2010.
4. EN 1264-2: 2008+A1:2012, IDT. Water-based surface-embedded heating and cooling systems. Part 2: floor heating: prove methods for the determination of the thermal output using calculation and test methods. European Committee for Standardization (CEN): Brussels, Belgium.
5. EN 15377-1:2008, ITD. Heating systems in buildings: design of embedded water-based surface heating and cooling systems. Part 1: determination of the design heating and cooling capacity. European Committee for Standardization (CEN): Brussels, Belgium.
6. Karpiesiuk P., *Research on the temperature field during surface heating*. Engineering Diploma Thesis, Faculty of Construction and Environmental Engineering, Department District Heating, Heating and Ventilation, 2015. (in Polish).
7. Żukowski M., Karpiesiuk P., Test results of a radiant radiator with a very low height. *Instal*, 10 (2015), pp. 38–41. (in Polish).
8. Karpiesiuk J., Chyży T., *Results of experimental tests on the temperature distribution of surface radiators of lightweight construction without screeds*. Rynek Instalacyjny, November 2017, No. 11/2017, year XXV, pp. 45–50. (in Polish).

9. Karpiesiuk J., A new dimension of screed-free heating and cooling (2). *Magazyn Instalatora*, 2 (234), February 2018, pp. 24–25. (in Polish).
10. Vademecum Elektra, Table 2.1.6, page 19, Poland, Edition 02/2024.
11. Rosiński M, Spik Z., How quickly does an underfloor heater cool down and heat up? Polish Installer, 2006.03., 8–17. (in Polish).
12. Radiant heating and cooling boards of the KAN-therm system. Poland, 03/2022. (in Polish).
13. Żukowski M., Karpiesiuk P., Test results of type B surface heating performed using dry technology. *Instal*, 7/8 (2015), pp. 33–37. (in Polish).
14. Roy B., *Multi-criteria decision support.* WNT, Warsaw, 1990. (in Polish).
15. Szwabowski J., Deszcz J., *Methods of multi-criteria comparative analysis.* Silesian University of Technology, Gliwice, 2001. (in Polish).
16. Ignasiak E., *Operational research.* PWE, Warsaw, 1996. (in Polish).
17. Ezersky V.A., Monasteries P.V., Merkusheva N.P., Instrumental approach to the assessment of consumer qualities of quartiers on the property market. *Scientific Proceedings of RAASN*, 2 (2017), pp. 169–180. (in Russian).
18. Nieto-Morote A., Ruz-Vilaa F., Fuzzy AHP multi-criteria decision-making approach applied to combined cooling, heating, and power production systems. *International Journal of Information Technology & Decision Making*, 10 (03) (2011), pp. 497–517. https://doi.org/10.1142/S0219622011004427
19. Greco S., Figueira J., EhrgottMultiple M., *Criteria decision analysis. State of the art surveys.* Springer, New York, NY, 2016. https://doi.org/10.1007/978-1-4939-3094-4

7 Usability and Economics of Lightweight Radiant Heating and Cooling System (LRH/CS)

Radiant heating offers many advantages and is therefore becoming increasingly popular, despite its higher investment costs compared to other heating systems. In recent years, the use of a particular type of radiant heating has increased, namely, lightweight radiant heating. This type is particularly interesting because its design (no screeds) allows for efficient heating and cooling of rooms. Numerous studies and calculations have confirmed that radiant heating and cooling are comfortable, ecological, and more efficient than traditional radiant heating with screeds. Furthermore, a multi-criteria decision analysis (MCDA) determined that the lightweight radiant heating/cooling system (LRH/CS) is the most economical solution, based on multiple criteria for evaluating eight types of radiant heating. The monograph includes calculation methods for the thermal efficiency of LRH/CS, both with lamellas and without heat-dissipating lamellas, and summarises the most important features. Numerical calculations and mathematical formulas clearly indicate that the observed error in the Scandinavian standard must be corrected to ensure accurate thermal efficiency results. The LRH/CS requires further research, and directions for its development are therefore outlined.

Radiant heating, and its most famous form of lying on the floor, is becoming increasingly popular worldwide due to the many advantages of this type of heating system. This type of heating with various types of screeds has been tested and proven in practical use. Scientists are still looking for other solutions to improve surface heating, adapting it to global trends, based, in particular, on the use of natural energy sources to counteract increasing environmental pollution and global warming. This can be achieved by installing devices that use renewable energy sources. Their use will be most efficient and useful when using natural energy resources at the lowest possible temperature. It seems that the LRH/CS meets all the necessary criteria. This publication proves, based on the extensively cited worldwide literature, the research performed, the author's experiments, and the development of a methodology for calculating this system without a heat-conducting layer, that the LRH/CS structure perfectly fits into the trends of global concern for ecology and is characterised by high heating efficiency (comparison to traditional capacities given in Tables 6.2–6.9, Chapter 6 of MCDA) and high comfort of use. This system seems to fulfil heating purposes and can be a very efficient source of cooling. It is possible because of its light and large surface design. This is even more important, because the demand for cooling has been increasing in recent years as the climate warms.

DOI: 10.1201/9781003679004-7

It would be worth continuing the work of Norwegian scientists, who were the only ones in the world who noticed the need to design lightweight radiant heating systems (LRHS) in 2001 by issuing a national standard [1]. It was then that they developed a methodology for calculating the thermal power and the maximum water supply temperature in the heating coil for lightweight underfloor heating systems without screeds and only with heat-dissipating lamellas. Other structures of this type of floor, e.g., without heat-dissipating lamellas, should be calculated based on individual experience. The continuation of their work is also important to eliminate the error in one of the formulas (noticed by the author of this monograph), which calls into question the correctness of the final result. This has also been noticed by other scientists, who compared the results obtained in calculations based on the Norwegian standard with other standards and numerical calculations. The topic is so interesting that the author of this monograph still has been waiting for the opinion of the representative of the Norwegian standard since mid-2023, after forwarding his comments questioning the correctness of one formula from the standard [1]. A correct response by the scientists who developed this Norwegian standard or their successors could shed new light on eliminating the likely error.

7.1 LRH/CS FEATURES CONFIRMED BY RESEARCH AND ANALYSIS

- Ensuring comfort in use when the assumed distances of the heating coils are up to 20 cm (when using heat-dissipating lamellas) and up to 15 cm (without lamellas arrangement).
- Higher thermal efficiency compared to all radiant heating systems with screeds – "wet" or "dry", according to [2], where the average floor surface temperature and heat transfer coefficient of the lightweight radiant floor system are 6.3% and 9.6% higher, respectively, than those of the traditional radiant system. The annual cost of the lightweight system is only 77.8% of the traditional cost.
- Possible installation, not only of lightweight heating with metal heat spreaders, but also without them, if we use smaller heating coil spacing (preferably up to 12.5 cm).
- Design and versatility of use horizontally and vertically allows for heating in winter and cooling in summer. The most universal place for this type of application is the building wall partitions. However, it is worth noting that a lightweight cooling system covered with a mineral hydrophobic mortar (e.g., gypsum or adhesive) has a better ability to avoid water condensation under identical ambient conditions. This can guarantee the safe and comfortable operation of the lightweight cooling system.
- LRH/CS technology provides better thermal comfort than convection cooling technology with lower energy consumption. The literature from this monography confirms that the energy savings from using a radiant system are up to 41% compared to those of an air conditioner. In addition, another similar study showed 34% energy savings compared to a traditional air conditioning system.
- Cost of the system is comparable to traditional underfloor heating.

- It is an ideal solution for devices using renewable energy sources, due to very low thermal inertia. To achieve a maximum floor temperature of 29°C, the sufficient supply temperature is 30°C if we use aluminium lamellas, or 36°C if we do not use them.
- The maximum calculated temperature feeding the heating coil, when installing aluminium lamellas, with a ceramic tile floor, cannot be higher than 42°C, or 51°C when covered by artificial floor coverings.
- When selecting thermal efficiency, one can use the calculations of the Norwegian standard [1], *only after correcting* the formula for calculating K_H (balanced heat transfer coefficient) following Chapter 4 of this publication, when the distance of the heating coil, placed on the lamellas, is no more than 15 cm. With the larger spacing of heating pipes (20 cm), standard calculations of thermal efficiency seem to be overestimated, and the supply temperature of 28°C and 35°C should be reduced by approximately 10%–20%, respectively. This was confirmed by experimental results.
- There is no standard formula for calculating the heat density flux when metal lamellas are not laid. The use of ready-made patterns of existing European standards that do not apply to lightweight radiant systems contributes to overestimating the thermal efficiency results by approximately 35% when the spacing of heating pipes is 12.5 cm. This results from the comparison of the LRHS calculation results without lamellas in this monograph. This publication develops its author's methodology for calculating the thermal efficiency q_{ob} of LRHS without lamellas – Section 4.3. Until standards for this application are issued, it is recommended to use the guidelines of the author of this monograph for q_{ob} calculations, but only to 15 cm between the distance of the pipe.
- After conducting an MCDA of eight different underfloor heating systems, it was analytically proven that "lightweight" underfloor heating with the use of aluminium lamellas is the most advantageous, and the next position was taken by the same heating system without the lamellas. Generally, it has been mathematically proven that the LRHS is the best solution for all radiant heating systems.

7.2 LRH/CS DIRECTIONS OF DEVELOPMENT AND RESEARCH

- It is worth having other scientists make their calculations taking into consideration the corrections of the Nordest NT VVS 127 formulas in this monograph. Calculations based on the modified formulas of the Norwegian standard (1) and (2) should be supported by experimental tests. Test models should be prepared following the Nordest standard without using any screeds. This will allow one to reproduce the actual operating conditions of lightweight underfloor heating without screeds. As a result, a comparison of calculations based on modifications to the formula from the Norwegian standard and calculations of experimental research should provide an answer as to the validity of the correction of the formula given by the author of this monograph.

- During subsequent experiments (preferably “in situ”), an effort should be made to develop a formula for calculating the thermal efficiency of lightweight underfloor heating without heat-dissipating lamellas.
- In future experiments, develop a formula for calculating the thermal efficiency of lightweight radiant heating of building wall partitions with and without lamellas.
- Future experiments (preferably “in situ”) should designate the heat transfer coefficient and develop cooling efficiency formulas in horizontal and vertical systems.
- A national and European standard should be developed for the dimensioning, design, and installation of LRH/CS.

ABBREVIATIONS

LRH/CS lightweight radiant heating/cooling system
MCDA multi-criteria decision analysis

REFERENCES

1. NORDTEST Method NT VVS127: UDC 697.1: Approved 2001-11. Floor heating systems. Design and type testing of waterborne heat systems for lightweight structures. Finland.
2. Liu M., Zhang H., Zheng W., You S., Heat transfer research of a new type of radiant floor heating system. *Heat Transfer Engineering*, 41 (2020), pp. 1626–1641. https://doi.org/10.1080/01457632.2019.1661701

Index

Note: **Bold** page numbers refer to tables; *italic* page numbers refer to figures.

A

Absence of screeds, 10, 21, 51
Accumulation system, 161, 167
The air temperature, 21, 26–27, 29, 30, 39–40, 46, *53*, 65, 77, 100, 101, 105, 121, 134
Aluminium foil, 12, *13*, *36*, 94, 100, 159–160
heat-conducting, 52
lamellas, **159**
Anhydrite screed, **159**
Arendal laminate, **97**, **141**
Author's, 70, **80**, 82–84, 188
calculation, 87–89
formula/s, 54
original calculation, 52, 81
own calculation, 78
own formula/s, 72
own methodology, 90, 190
Average flooring temperature, 52

B

BAUNORM Fibercement, **114–116**, **127–129**, **146**, 147, **148**, 153
board, 98–101, **98**, 116–117
dry screed, 113, 127
screed, 146–147, 149, 154

C

Cable, 19
heating, 12, *14–17*, 19, 53, 83, 159–161, 167–169, 171
Calculations of the thermal output, 94, 101
Cement screed, *15*, *27*
dry, **159**, 160–161, 163–165
wet, 21, 51, 159
Central zone, 46, 63, 71, 76, 85, 87, 106–107, 121–122, 135–136
calculated for, 38
and upper, *68*, *80*
Ceramic tile flooring, 51, 72, 74, *See also* Ceramic tile floors
Ceramic tile floors, 51, 94, 144
Coil, *7*, *27*, 40, *53*, **87–88**, 90, 95, 101
covered, 21, 51
with a diameter, 83, 161
different, 28, 102, 108, 112, 117–118, 124, 130–131, 138, 143, *150–152*
distance, 59, 155
placed, 52, **161–167**, 190
spacing, *27–28*, *43–44*, *67–68*, 89–90, *150–152*
supply temperature, 52, 56, 64, **67**, 70, 76, *80*, **81–83**, **114–116**, **127–129**, **141–148**
using, 21, 51
wall, 58
Cold spots, 95, 149
Comfort of use, 85, 90, **162–166**, **169**, 173, 188
improve, 38, 158, 185
Comfortable temperature distribution, 52, 89
Comparative analysis, 95, 158, 181, 183
Completion rate, **162–169**, *See also* Pace of installation
Completion time, **162–169**, 181, *See also* Construction time
Configurations of floors, 94
Construction time, 172, 176
Criteria graph, 175–176, *176*
Criteria measures, 177, 181, **182**

D

Denser coil spacing, 21, 51
Description of criteria, 171
Dissipating, 161, 163
layer, 102, 117, 130
Dry screed, 2, 36, 55–56, 84, **98–99**, 101, 117, 119
board, 99
cement screed, **159**
Conductive/conducting layer, 119
floor without a, 140
in the heated floor, 19
heavy, **83**
and metal lamellas, 111, 113
slab, 100
with steel lamellas, **183**
thick/thin, 42
types, *37*, 130
under, 46
and/with aluminium lamellas, 108, 113, 155, **183**
without aluminium lamellas, 127, **183**

E

Ease of installation, 98, 162, 173
numerical rating, **162–169**

Edge zone, 63, 76, 86, **88**
in the edge zone, 22, 38, 46, 61, *68,* 73, 106–107, 122–123, 135–136
with a heating coil, *68*
upper border/limit, *80*
Efficient, 1, 29, 32, 51, 91, 188
Electric floor heating, 12, 19, 167
Electric heating, 1, 19, **167–169,** 185, *See also* Electric floor heating
cable, 52, 83–84, **159–160**
with heating cables, 161, 167
with heating mats, 161, 167, 185
"light"/lightweight, 12, 18
mat, *18*, **159–160, 183**
radiant, 12
type of, 12–13
Energy-efficient buildings, 21, 47, 52
Environmental protection, 24–25, 92, 94
Equipment, 24, 91, 154, 162
Essential parameter, 95
Eternal PVC, **97, 142**
Experimental calculations, 36, 52
Experimental studies, 51, 55, 87, 90, 95–96, 145, 155

F

Fibre mesh, 36, 153, 161
glass, 153, 161
Finishing layer, *2*, 95, 130
Floor heating
electric, 12, 19, 167
Floor temperature, *15*, 45, 53, 78, 170, 173
average, *43–45*, 56
comfortable, 22
higher, 29
maximum, 91, 190
optimal, 21, 46, 51
permissible, 22
Flooring or coverings, 94
Flotex colour carpet, **97, 115, 129, 142**
Functional values, **162–169,** 173

G

Gypsum screed, 82, **159,** 160–161, **163–164**

H

Heat diffusers, 10, 38, 46
without heat dissipation, *9*, 38, 91
Heat-dissipating elements, 51, *See also* Heat-dissipating lamellas
Heat-dissipating lamellas, 80, 89, 94, 143, 145, 155
aluminium, 95, 100, 140
using, *6*, 189
with, 51, 64, 81, 189
without, 52, 76, 80, 96, 161, 163, 188–189, 191
Heat emitter, 95
Heat flux density, 26–27, 38, 55–56, 65, 79, 89–90, 101, 150
calculating/calculation, 55, 95, 101, 108, 116, 124, 130, 138
determine the, 29
at different coil spacing, *28*
at different room temperature, *27*
emitted from, 55
experimental, 65, 77–78
high, 100
improving, 100
lower, 45
maximum, 55–56, 58–59, 63, 70–72, 75, 101, 107, 117, 122, 130, 136
output, 94–95
of the radiator, 30
theoretical, 55, 81, 154
thermal/heat efficiency, 56, 72, 76, 173
thermal power, 55, 64, 70, 100, 116, 143
value, 88, 95
Heat spreaders, 26, 53, 84, 87, 89–90, 189
Heating coil, 38, 45, 47
spacing, 40, 42, *43–44*
Heating/cooling, 30–32, **35**
radiant, 10, 26, 33, **35**, 40, 90, 188
surface, 26, 170
system, 30, **31**, 99, 187
underfloor, 47
wall, **35**
Heating foil, *13*, 13
Heating mat, 12, *13*, 13, 16–*18*, 19, 159–161, 167, **168–169,** 185
Heating pipe/s, *1–4, 9*, 97–99, 101, 111, 117, 130, 155
above, 31, 85
below, **31**
between, 40, 95
construction, 91
diameter of the, 57, 71, 73
distance, 83, 88
dry, 82, **83**
electric, 17
electric underfloor heating, 17
heavy, 10, 12, 19, 30, 82, 162
higher heat power, 21, 51
medium in the, 67–68, 70, 112, 114
screeds, 12, 46, 95
spacing, *2,* 45, 57–59, 63, 71–72, 75, 84
system, 1, 21, 27, 46, 52
temperature of/on, 53
traditional surface, 30, 46
underfloor heating, 27–28, 83, 160
wall, 59
water system, 46
wet, 82, **83,** 84

I

Impact, 29, 34, 95, 144–145, 147, 149
airflow, 29
of each layer, 154
on the environment, 35
negative, 24–25, 35
significant/ly, 29, 45
small, 29, 94
on thermal efficiency/output/power, 45, 94, 143, 145, 154
of thermal insulation, 153–154
zone, 105, 107, 121–122, 134, 136
Insulation board, *9*, 97, 102, 117, 130, 140, 143, 153

K

Karpiesiuk and Chyży, 6, 30, 52, 87, 161
Karpiesiuk J., 31, *67–68, 80, 85–86,* 87, 95–96, 155, 161, *See also* Karpiesiuk and Chyży
Knauf brio, *98,* 113, **114–116, 127–129, 146, 148,** 153
gypsum-fibre board, 98–99
Knauf dual floor, **98**, **114–116, 127–129, 146, 148,** 153–154, **159–160**, **163–164**
dry screeds, *36–37,* 113, 127, 146, 153
were laid, 97
plasterboard, 99
thin, 94, 96, 98, 143
three, **98–99**
under, 46
with or without lamellas, 84, 95
without, 95, 116

L

Laminated panels, **144**, **146**, 147, **148**, 154–155
Lightweight flooring installation, 95
Lightweight system, 5, 10, 83, 90, 160, 189
Load-bearing layer, 94, 96
Low-temperature, 29, 31, 173

M

Manufacturer's warranty, **162–169**, 173, 176
Medium temperature, 47, 91, 111, 143, 152–153, 155
Mesh layer, *18*, 38
adhesive, 38, 94, 96–97, 130, 134, 138, 153–154, 163
Metal dissipative layers, 52
Metal heat-dissipating elements, 51
Metal lamella, 21, 47, 51, 57, 96, 99–101, 143, 145, 149, 190
dry screed and, 155
heat-dissipating, 10, 116, 130, 143
screed and, 111, 113
with, 99
without/no, 95, 99, 116, 129, 147, 154–155
Microflex
microcement, 114, 127, 140, **144, 146, 148**
microflex two-component, **97, 115, 128, 142**
Models with lamellas, 51, *67*
Most effective, 97
on the ceiling, 21, 51
on the floor, 21, 51

N

Natural energy resources, 188
Net cost per 1 m^2, **162–166**, **168**
No screed, 10, 51, 95, 129, 153, 188
Norwegian standard, 10, 55, 81, 88–90, 95–96, 189–190
The NT VVS 127:2001, 51–52, 70, 96
Nordest, 55, 59, 67, 72, 190
Nordest method, 56
standard, 54, 81
Numerical calculations, 42, 52, 56, 83–84, 95–96, 188–189
of a "light" floor, 46
of an LRHS, 45
of thermal power, 52

O

Operating cost, 19, 23, 153, 155, **163–168,** 174, 179, 182, 184–185
Output data, 143, 145, 147, 161

P

Pace of installation, 172
Placed in the C2S1, 161
Placed in thermal insulation, *2*, 161, 163, **163–167**
PVC floor panels, **144**, **146, 148**
vinyl, **97, 114, 127, 141**

R

Radiant heating, 46, 188
benefit/s of, 29
can reach, 29
or cooling, 24, 40, 90, 188
electric, 12, 19
installed, 21, 51
lightweight, **34,** 51, 91, 94, 188–189, 191
models, *44*
system, 1, 19, 26, 32, 39, 52, 94, 189–190
thin, 91
wall, 31–32, 42
with, 42
without, 5, 52

Radiant heating and cooling system/structure, 4, 10–11, 25–26, 30, 33, 47, 90–92
advantage/s, 10, 30, 46, 91, 96
of a lightweight, 38, 47
many advantages, 10, 25, 51, 188
of wall radiant, 31, 42
Renewable energy sources, 21, 25, 29, 38, 91, 188, 190
integration with, 52
Retail/wholesale prices, 160, 167

S

Scandinavian standard, 100, 116, 130, 188
Screed layer, 1, *2–4*, 10, 96, 119, 147, 153
self-levelling, 129
Screedless, 46, 51, 153
Selected criteria, 173, 178
Self-levelling, *16,* 20, 130
cement layer, 98
mortar, 16–17
screed/s, 94, 96–97, 99–100, 129, 132, 154
Set of criteria, 171, 175–176, **183,** 184–185
Sikafloor elastic resin, **97, 116, 129, 142**
Solution for radiant heating, 5, 52
Surface heating, 10, 29, 34, 36, 94, 188
and cooling, 1, 10, 21, 32, 47, 155, 170
electric, 12, 19
lightweight, 21, 32, 36, 51–52, 84
under the plasterboard, 31
system/s, 21, 51, 94
Surface radiator, 5, 21, 40, 170, 184
was calculated, 65, 77
light/lightweight, *6*, 30, *38*, 84, 88, 167
ultra-thin, 36, *37,* **88**, 91
wall-mounted, *32*, *80*

T

Theoretical and experimental methods, 52
Thermal efficiency calculations, 51, 95
Thermal inertia, 10, 162, **162–169,** 170, 175, 183
of the floor, 173
of the heating system, 172, 176
high, 46
low, 31–32, 35, 38, 46, 100, 172, 190
surface, 21, 51
Thermal output, 47, 99–101, 130, 143–147, 149, 151–155, *See also* Heat flux density
calculated, 143, 145, 149, 154
calculating, 10, 145, 147
of ceramic tile, 94
and comfort values, 90
efficiency, 94
heat flux density, 94
higher/highest, 143–144, 146, 149, 154
increases, 143, 147, 153
largest, 145
of lightweight heated floors, 94, 99
lowest, 144, 146, 149
maximum, 143, 145, 147
result, 90, 94, 154–155
value, 94, 130, 144–146, 149
Thermal power calculations, 114, 127
theoretical, 52, 155
Thermal resistance values, 31, 95, 104, 119, 133
Thin, dry screeds, 94, 96, 98
Thin screeds, 95
with, 100, 114, 116, 143
without, 94
Three-layer wooden board, **97,** 113, **114,** 127, **128,** 140, **141, 144, 146, 148**
Total investment cost, **162–169**, 172
Traditional, 4, 29, 31, **162**, **183**, 184–185, 188–189
heating system, 21, 90
"heavy", 46
surface, 21, 30, 52
underfloor/radiant heating, 51, 188–189
water heating, 161–162
"wet", 160

U

Underfloor heating. *See* Floor heating
Underfloor heating with, 21, *43*, 55–56, 88, 95, 162–167, 190
Underfloor heating without, 42, 51, 81, 90, 190–191

V

Various coil spacings, 95
Various types
of buildings, 94
of conductive panels, 51
of dry screeds, *37,* 188
of facilities, 92
of heating floors, 154
of heavy screeds, 12
of flooring, 12, 94
of LRHS, 95
of metal, 1, 12
of stone tiles, 95
of underfloor heating, 83
Versatility, **162–169**, 183, 185, 189
of using LRH/CS, 91

W

Wall heating, *6*, 19, 30–31, 47
and cooling, 30, 33, **35**
surface/radiant, 34–35, 47
Water heating, 12–13, 19, 161–163, 184–186
and cooling, 6

Waterpanel E, *80,* 100, 102, 117, 130
Waterpanel X, *80,* 100, 102, 117, 130, **159–160,** 160
Waterpanel XE, **159–160**, 160
Weight of
  1 m^2, **162–169**
  320g/m^2, 160
  the heating system, 172, 175
  the i-th criterion, 177, **179,** 180
  materials, 162
  the screed, 172
Werner-Juszczuk, 27, 42, 52, 55
Wet cement screed, 21, 51, 160
"wet" concrete, 161–162, 167
Wet screed, 13, *14,* 16–17, 31, **83,** 84, 96, **159–160, 162**
  concrete, 167, **168**

Z

Żukowski, 19, 36, *39,* 170
  and Karpiesiuk P., 38, 161, 170

For Product Safety Concerns and Information please contact our EU representative GPSR@taylorandfrancis.com
Taylor & Francis Verlag GmbH, Kaufingerstraße 24, 80331 München, Germany

www.ingramcontent.com/pod-product-compliance
Ingram Content Group UK Ltd.
Pitfield, Milton Keynes, MK11 3LW, UK
UKHW021955111026
476504UK00003B/22
* 9 7 8 1 0 4 1 1 5 3 5 2 8 *